The Blue Economy in Sub-Saharan Africa

The blue economy, comprising coastal and marine resources, offers vast benefits for sub-Saharan Africa: of the 53 countries and territories in the region, 32 are coastal states; there are 13 million sq km of maritime zones; more than 90% of the region's exports and imports come by sea; and the African Union hails the blue economy as the 'new frontier of African renaissance'.

Despite their importance, the region's coastal and marine resources have been neither fully appreciated nor fully utilized. They are only now being recognized as being key to Africa's potential prosperity. As the region grows, it has, in general, not taken adequate safeguards to protect these valuable resources. That is partly because some of the problems (pollution, for example) are regional and know no borders. All too often, short-term gains are made at the expense of the long term (overfishing, for example).

This book provides, for the first time, a study of the constraints and opportunities the blue economy offers for sub-Saharan Africa. It includes an introduction and overview; sectoral analyses (including tourism, fisheries, mineral resources, culture, shipping and maritime safety); country case studies; and analyses of regional and international efforts towards better coastal zone and marine management.

Donald L. Sparks, PhD, is Emeritus Professor of International Economics at the Citadel in Charleston, South Carolina, USA, and a university lecturer in International Business at the Management Center Innsbruck, Austria.

Europa Regional Perspectives

Providing in-depth analysis with a global reach, this series from Europa examines a wide range of contemporary political, economic, developmental and social issues in regional perspective. Intended to complement the Europa Regional Surveys of the World series, Europa Regional Perspectives will be a valuable resource for academics, students, researchers, policymakers, business people and anyone with an interest in current world affairs with an emphasis on regional issues.

While the Europa World Year Book and its associated Regional Surveys inform on and analyse contemporary economic, political and social developments, the Editors considered the need for more in-depth volumes written and/or edited by specialists in their field, in order to delve into particular regional situations. Volumes in the series are not constrained by any particular template, but may explore recent political, economic, international relations, social, defence, or other issues in order to increase knowledge. Regions are thus not specifically defined, and volumes may focus on small or large group of countries, regions or blocs.

The Caribbean Blue Economy
Edited by Peter Clegg, Robin Mahon, Patrick McConney and Hazel A. Oxenford

The Future of the United States-Australia Alliance
Edited by Scott D. McDonald and Andrew T. H. Tan

External Powers in Latin America
Geopolitics Between Neo-extractivism and South-South Cooperation
Edited by Gian Luca Gardini

The Blue Economy in Sub-Saharan Africa
Working for a Sustainable Future
Edited by Donald L. Sparks

For more information about this series, please visit: www.routledge.com/Europa-Regional-Perspectives/book-series/ERP.

The Blue Economy in Sub-Saharan Africa

Working for a Sustainable Future

Edited by
Donald L. Sparks, PhD

LONDON AND NEW YORK

First published 2021
by Routledge
2 Park Square, Milton Park, Abingdon, Oxon OX14 4RN

and by Routledge
605 Third Avenue, New York, NY 10158

Routledge is an imprint of the Taylor & Francis Group, an informa business

British Library Cataloguing in Publication Data
A catalogue record for this book is available from the British Library

Library of Congress Cataloging-in-Publication Data
Names: Sparks, Donald L., editor, author.
Title: The blue economy in Sub-Saharan Africa : working for a sustainable future / edited by Donald L. Sparks.
Other titles: Europa regional perspectives.
Description: New York : Routledge, 2021. | Series: Europa regional perspectives | Includes bibliographical references and index.
Identifiers: LCCN 2020056356 (print) | LCCN 2020056357 (ebook) | ISBN 9780367422127 (hardback) | ISBN 9780367822729 (ebook)
Subjects: LCSH: Marine resources development–Africa, Sub-Saharan. | Marine resources–Economic aspects–Africa, Sub-Saharan. | Coastal zone management–Africa, Sub-Saharan. | Sustainable development–Africa, Sub-Saharan.
Classification: LCC GC1023.88.A357 B58 2021 (print) | LCC GC1023.88.A357 (ebook) | DDC 333.91/640967–dc23
LC record available at https://lccn.loc.gov/2020056356
LC ebook record available at https://lccn.loc.gov/2020056357

ISBN: 978-0-367-42212-7 (hbk)
ISBN: 978-10-320-3456-0 (pbk)
ISBN: 978-0-367-82272-9 (ebk)

Typeset in Times New Roman
by Taylor & Francis Books

Contents

Illustrations

Figures

Tables

Foreword

Our distinguished quality as human beings has mainly been ascribed to our adaptability and creativity. It is therefore unsurprising that the blue economy concept is becoming increasingly popular and has gained considerable momentum in recent years. The blue economy is a principle that should come naturally to us as it makes use of our best qualities as human beings while simultaneously ensuring our continued existence and that of our planet.

Thirty-two nations in sub-Saharan Africa have coasts and some 90% of the region's trade is conducted by sea. Maritime zones under African jurisdiction total over 13 million sq km with vast potential for offshore and deep-sea mineral exploitation. Clearly coastal and marine resources will play key roles as a source of food, energy and economic development in the foreseeable future. To better understand the challenges facing these coastal nations, this volume (written by a cadre of world-renowned experts) will be an outstanding resource. It examines a variety of important issues such as fisheries and aquaculture, shipping, tourism, maritime trade, culture, energy, illicit trade, oceans governance and case studies from several African nations.

For us Small Island Developing States, and indeed other coastal states, the ocean is an intrinsic part of our livelihood and culture. As such, transitioning towards a blue economy is the most sensible choice, but it is not necessarily the easiest. In practice the implementation of this concept is contextual and country-specific, as there is no one-size-fits-all approach. While the task at hand may be daunting, as with many endeavours the load normally lessens with a helping hand.

The most important thing to acknowledge is that regardless of our geographical status as island, coastal or landlocked states, we are all subject to global connectivity. All of us are to a certain extent bound by both the positive and negative effects of our interconnectedness. For instance, more than 80% of the world's trade occurs via sea freight, and seafood with its high nutrition and protein value is consumed all around the world regardless of physical location. On the other side of the coin, as could be illustrated by Seychelles' pristine and seemingly remote island of Aldabra, no corner of the world is safe from ocean pollution. Due to our significant dependence on the ocean, and without being oblivious to the correlation between ocean health and climate change, it is evident that the impact of degraded ocean health will be just as far reaching.

As citizens of the world, we are therefore obligated to maximize our efforts to safeguard the integrity of our oceans, seas and fresh water sources against the many existing environmental, maritime and food security threats. This book will help us in this quest.

The blue economy is the key; we need to stand united and favourably mould our destiny!

Vincent Meriton
Former Vice-President of Seychelles

Acknowledgements

This volume was written at the height of the COVID-19 pandemic and the authors have done a remarkable job in getting their chapters completed in a timely manner. I would like to thank each of them for their professionalism, expertise and willingness to 'go the extra mile' for this project. Of course, the pandemic also made it difficult to offer more precise projections for the future. I would especially like to thank Cathy Hartley and Iain Frame at Routledge for their invaluable assistance.

The blue economy in sub-Saharan Africa has become increasingly important now that coastal states are seeking ways to strengthen their economies in order to improve livelihoods as well as to maintain safeguards for their fragile coastal and marine environments. We hope that in some small way this volume will move the discussion forward in a positive manner.

Donald L. Sparks, PhD
Charleston, South Carolina, USA

Contributors

Donald L. Sparks is Emeritus Professor of International Economics at the Citadel in Charleston, South Carolina, USA, and a university lecturer in International Business at the Management Center Innsbruck, Austria. He served as a staff assistant to US Senator Ernest F. Hollings and as a senior economist for the South Carolina Sea Grant Consortium where he directed its Indian Ocean Initiative. Prior to that he was the regional economist for Africa in the US Department of State's Office of Economic Analysis. He has received Fulbright awards in Laos, Slovenia, Swaziland (now Eswatini) and at the African Union in Addis Ababa, Ethiopia. He also chaired the Economics Department at the American University in Cairo, Egypt, for a year.

Dr Sparks has been a consultant writing on African economic issues for the Economist Intelligence Unit, the United Nations Industrial Development Organization and the United Nations Council for Namibia. He organized a National Science Foundation workshop in Mauritius on Ocean and Coastal Development and Protection and has written extensively on blue economy issues in the western Indian Ocean, visiting over a dozen states in the region.

Dr Sparks serves on the International Union for Conservation of Nature (IUCN) Commission on Environmental, Economic and Social Policy. He has authored the 'Economic Trends' chapter in *Africa South of the Sahara* (Routledge) annually since 1986. He holds a BA from the George Washington University, USA, and an MA and a PhD from the School of Oriental and African Studies, London, UK. Dr Sparks and his wife, Dr Katherine Saenger, live in Charleston, South Carolina, and Seefeld in Tyrol, Austria.

Jeff Ardron is an Adviser on Ocean Governance in the Trade, Oceans and Natural Resources Directorate of the Commonwealth Secretariat, London, UK. He was the main architect of the Commonwealth Blue Charter, an agreement adopted by Commonwealth Heads of Government in 2018. Dr Ardron has more than 30 years' experience in marine governance, planning, science and conservation, and has worked in a wide variety of sectors – governmental, inter-governmental and non-governmental, private, research and academia. In addition to several papers in the peer-reviewed literature, he has co-authored a textbook on international marine policy. Dr Ardron holds a PhD from the University of Southampton (Ocean and Earth Sciences), UK, and an MSc in Environment and Management, from Royal Roads University, Canada.

Dominique Benzaken is a policy specialist in blue economy, blue finance and ocean governance at the Australian National Centre for Ocean Resources and Security. She is a Visiting Fellow at the School of Public Policy within the College of Asia Pacific at the Australian National University, Australia, a senior adviser to the Ocean Assets Institute and the Global Island Partnership and she is a member of the IUCN World Commission on Protected Areas. Recent assignments and collaborations include financing mechanisms for the blue economy in Mozambique, regional sustainability governance for the Red Sea and options for a Pacific Ocean Blue Bond. As a former Commonwealth-funded expert over a period of two years she advised the government of Seychelles on blue economy and ocean governance. She was previously a senior marine policy adviser with the Nature Conservancy, the IUCN and the Secretariat of the Pacific Environment Programme, following a formative career with the Australian government and the Great Barrier Reef Marine Park Authority. She has a Bachelor's degree in Biological Sciences and a Master's degree in Tropical Ecology from James Cook University, Australia. She is a PhD candidate on sustainable blue economy at the Australian National Centre for Ocean Resources and Security, University of Wollongong, Australia.

Rosemarie Cadogan is an international lawyer and Legal Adviser in the Trade, Oceans and Natural Resources Directorate of the Commonwealth Secretariat, London, UK. Her public policy, project and transactional experience in negotiations, ocean and environmental governance frameworks and human rights spans over 20 years in Guyana, the Caribbean, West Africa, the Pacific and Canada. She holds a Master's degree in Marine and Environmental Law from Dalhousie University, Canada, a postgraduate diploma in International Relations from the University of Guyana and a Bachelor's degree in Law from the University of the West Indies. She is eligible to practise in Guyana, the Caribbean and Canada.

Charles S. Colgan is Director of Research for the Center for the Blue Economy at the Middlebury Institute of International Studies at Monterey in Monterey, California, USA. He is also Editor-in-Chief of the *Journal of Ocean and Coastal Economics* and an adjunct professor of International Environmental Policy. He is Professor Emeritus of Public Policy & Planning in the Muskie School of Public Service at the University of Southern Maine, USA, where he chaired the Graduate Program in Community Planning & Development. He served for 12 years in the Maine State

Planning Office including positions as Maine State Economist and Director of the Maine Coastal Program. He received his BA from Colby College and his PhD in Economic History from the University of Maine.

Vivian Louis Forbes is a Research Professor at the National Institute for South China Sea Studies, Hainan, and is affiliated with Wuhan University, China, the National Defense University, Malaysia, and the University of Western Australia. He is a professional practising cartographer, a marine political geographer, a lecturer in spatial sciences and maritime affairs, and a former merchant navy officer. He has developed expertise in international law particularly as it relates to maritime and terrestrial political boundary determination, law of the sea and associated issues; he has practical experience in terrestrial and hydrographic surveying; and he lectures in Law of the Seas issues, spatial sciences and political geography. He has an excellent understanding of the problems of maritime boundary determination, especially in the regions of South-East and East Asia and those ocean areas surrounding the Australian continent. He has authored several books, atlases, chapters in books and academic research papers.

Nick J. Hardman-Mountford is the Head of Oceans and Natural Resources at the Commonwealth Secretariat, where he leads a technical advisory team delivering the Commonwealth Blue Charter initiative and technical assistance projects on ocean governance and sustainable natural resources development in Commonwealth countries. Prior to joining the Commonwealth Secretariat, Dr Hardman-Mountford was principal scientist at Australia's national research agency, the Commonwealth Scientific and Industrial Research Organisation. His research on ocean observations and modelling has taken him to the Atlantic, Indian, Pacific and Southern Oceans and has found application in addressing a diverse range of societal drivers such as sustainable blue economies, carbon storage, food security, energy and climate change. Dr Hardman-Mountford has published extensively in international journals, books and technical reports. He is an Adjunct Professorial Research Fellow at the University of Western Australia.

Abdi Fatah Hassan is a Juris Doctor student at the University of Western Australia, where he also holds Bachelor's and Honours degrees in Socio-Legal Studies and History. His is also the recipient of the Law School award for International Environmental Law. Abdi has previously published research on the role of public international law in the processes of (de)colonization in the mid-twentieth century. His interests include the large-scale environmental effects of climate change on vulnerable populations in the global South, and the functions of international environmental law in addressing these vulnerabilities.

Kelly Hoareau is the Director of the James Michel Blue Economy Research Institute at the University of Seychelles. She is currently developing academic resources, networks and knowledge to support the Seychelles' Blue

Economy Roadmap. Internationally, she also contributes to research, education and awareness to advance ocean-based sustainable development. Kelly has first-hand experience of working in Africa and with various small island developing states and costal nations, across marine and terrestrial ecosystems. Coupled with her engagement with academic, government, private and non-governmental entities, this has led to her interest in multi-partner projects facilitating co-developed knowledge and stakeholder engagement that support sustainable development and environmental management agendas. She is involved with various transdisciplinary projects (local and international) and publications related to the aforementioned subject areas. Kelly holds undergraduate degrees in Environmental Management and Geology, as well as a postgraduate degree in Environmental Management from the University of Johannesburg, South Africa.

Angela M. Lamptey (aka **Angela M. Ahulu**) is a lecturer in the Department of Marine and Fisheries Sciences at the University of Ghana. She has a PhD in Fisheries Science from the University of Ghana. Her research interests include aquaculture, fisheries biology and ecology, biodiversity conservation (seabirds and sea turtles), aquatic biotopes and water quality studies. She publishes in and is a reviewer for a number of international peer-reviewed journals. Currently, her research areas include the ecology of aquatic biotopes, the food and feeding habits of selected marine and freshwater fishes, the water quality assessment of coastal lagoons in Ghana, studies on ecotoxicology in seabirds, and child labour in the fishing industry. She is affiliated to the Ghana Science Association, the Society for Conservation Biology and the Royal Society of Chemistry. She is a fisheries and ornithology consultant to two non-governmental organizations, namely the Centre for Maritime Law and Security and Environmental Solutions Ltd. She was recently nominated as an independent expert by the Ad Hoc Expert Group for the African Union Commission which is tasked with drafting the annexes for the finalization of the Lomé Maritime Charter. She is one of the quiz mistresses for the National Science and Maths Quiz, is a motivational speaker to various youth groups and a career guidance adviser to secondary school girls. She is passionate about human rights issues.

Jade Lindley has a PhD from the Australian National University and is a criminologist at the University of Western Australia Law School. Jade specializes in transnational organized crime and the intersection with international law, and has a particular interest in maritime crimes especially maritime piracy and illegal fishing in the Indo-Pacific. Jade's PhD applied criminological theory to Somali piracy to understand the most appropriate international and regional responses to bring it under control. It was published in 2017 as a monograph entitled *Somali Piracy: A Criminological Perspective.*

Clever Mafuta is the Waste and Marine Litter Head of Programme at GRID-Arendal. Clever has over 20 years' experience in environmental assessment and reporting. His academic background is in agriculture, and he has a Bachelor of Science (Honours) degree in Agriculture and a Master's degree in Crop Protection, both from the University of Zimbabwe. Clever also studied for a Master's in Business Administration at Nottingham Trent University, UK. Notable career achievements include being one of the researchers and writers of *Southern Africa Environment Outlook* (2008) and the lead author of *Zambezi River Basin: Atlas of our Changing Environment* (2012) and *Limpopo River Basin: Changes, Challenges and Opportunities* (2017). In his professional capacity Clever has co-led regional and global environment assessment processes that include the sixth edition of the *Global Environment Outlook* report; the fifth edition of the *Global Environment Outlook* report (GEO-5); co-coordinating lead author for Chapter 6 (Regional Perspectives) of the fourth *Global Environment Outlook* (GEO-4) report; and coordinating lead author for the third *Africa Environment Outlook* report.

The Honourable Vincent Meriton (Vincent Emmanuel Angelin Meriton) served as the Vice-President of Seychelles from 2016–20 and was responsible for Information, Blue Economy, ICT, Industries and Entrepreneurship Development, Religious Affairs and Civil Society, and Inner and Outer Islands. The Department of Foreign Affairs was assigned to Vincent in 2018 He served as Seychelles' Minister for Community Development, Social Affairs, Sports, Minister of Social Affairs and Employment, and Minister of Labor and Social Policy. He graduated with a Master's degree in Sociology from the Moscow State University, Russia.

Francis A. Mwaijande is a Fulbright Scholar in Public Policy from the University of Arkansas, USA, where he received his PhD. He holds a BA in Education from the University of Dar es Salaam, Tanzania, and an MA from the University of Wolverhampton, UK. He is a senior lecturer in Public Policy and Social Science Research at Mzumbe University, Tanzania. In 2008 he was an adjunct instructor in the Department of Political Science, University of Arkansas, USA, where he taught Comparative African Politics focusing on social policy reforms including public sector reforms, democratic reform process and the United Nations Millennium Development Goals. He was a principal investigator for the Food Price Trends Analysis for Policy Options for Enhancing Food Security in Eastern Africa project of the Association for Strengthening Agricultural Research in the Eastern and Central Africa. Dr Mwaijande contributed to the Assessment Development Results for UNDP Tanzania. He is the chairman of the Tanzania Evaluation Association, working on building capacity for conducting evaluations, creating culture for demand and use of evaluations in government and parliament. He published the *Handbook on Evaluation Research Methods and Practice* (2012).

Iddi R. Mwanyoka is a lecturer at the University of Dodoma, Tanzania. He holds a Master's degree in Water Resources Management and a postgraduate diploma in Poverty Analysis for Socio-economic Security and Development. He has conducted a wide range of research and consultancy and has published a number of papers in these areas.

Chilenye Nwapi is a Legal Adviser in the Trade, Oceans and Natural Resources Directorate of the Commonwealth Secretariat, London, UK. He supports the implementation of the Commonwealth Secretariat's technical assistance programme on petroleum and mining (both deep-sea and land-based) reform in Commonwealth member states. Dr Nwapi holds a PhD in Law (specializing in resource extraction, business and human rights) from the University of British Columbia, Canada, and an LLM in Natural Resources and Environmental Law from the University of Calgary, Canada, where he later held the Banting Postdoctoral Fellowship. He is a barrister and solicitor of the Supreme Court of Nigeria.

Thean Potgieter is an Associate at the Wits School of Governance, Witwatersrand University and the Centre for Military Studies, Stellenbosch University, South Africa. Professor Potgieter is Chief Director of Research and Innovation at the South African National School of Government. Previous experience includes Director of the Centre for Military Studies, Departmental Chair of Military History (Stellenbosch University) at the School for Security and Africa Studies, as well as service in the South African navy. He has vast teaching and research experience, has acted as a guest lecturer and research fellow in a number of countries and has published books, chapters in books, articles and position papers internationally. Recent publications include an edited volume, *Reflections on War: Preparedness and Consequences* (2012), various articles on maritime affairs, and three chapters in a volume on *African Military Geosciences* (2007). He is a member of the South African Academy for Science and Art and is Chief Editor of the *Africa Journal of Public Sector Development and Governance*. Professor Potgieter holds degrees in the Human Sciences, History and Strategic Studies from the University of Johannesburg, as well as a DPhil from Stellenbosch University.

Alison Swaddling is an Adviser on Ocean Governance in the Trade, Oceans and Natural Resources Directorate of the Commonwealth Secretariat, London, UK. She provides direct assistance to Commonwealth governments through her work on governance and policy development, has developed regional frameworks for marine scientific research and environmental management for deep-sea minerals, and leads the Commonwealth's deep-sea minerals project. Alison has previous experience in the commercial minerals sector and as a guest lecturer at the University of the South Pacific. She holds a MA in Environmental Law from the University of Sydney and a BSc in Marine Science from James Cook University, Australia.

Erika Techera is a Professor of Law at the University of Western Australia. She holds an LLB with Honours, Master's degrees in Environmental Law and International Environmental Law, and a PhD in Law. Erika specializes in international and comparative environmental law, particularly ocean governance in the Indo-Pacific region. Erika is the author of over 100 publications addressing issues such as marine protected areas, fisheries regulation, marine pollution law, safeguarding cultural heritage, marine-based tourism and the conservation of marine megafauna (particularly sharks). She also has a specific interest in multidisciplinarity, including the interface of law, science and governance. Prior to becoming an academic, Erika practised as a barrister and has also been a company director. She is a Fellow of the Australian Academy of Law.

Torsten Thiele is an ocean governance and finance expert based in Berlin, Germany. He is the founder of the Global Ocean Trust, a Senior Research Associate at the Institute for Advanced Sustainability Studies in Potsdam, Germany, a Visiting Fellow at the LSE Institute of Global Affairs, a Senior Advisor to the IUCN Blue Natural Capital Financing Facility and a member of the Scientific Committee of the Ocean-Climate Platform. He can draw on over 20 years' experience in project and infrastructure finance with leading finance institutions. Recent publications addressed blue infrastructure and nature-based solutions, innovative High Seas finance, blue bonds, deep-sea governance and ocean climate finance. Torsten holds graduate degrees in Law and Economics from Bonn University, Germany, an MPA from the Harvard Kennedy School, USA, and an MPhil in Conservation Leadership from the University of Cambridge, UK.

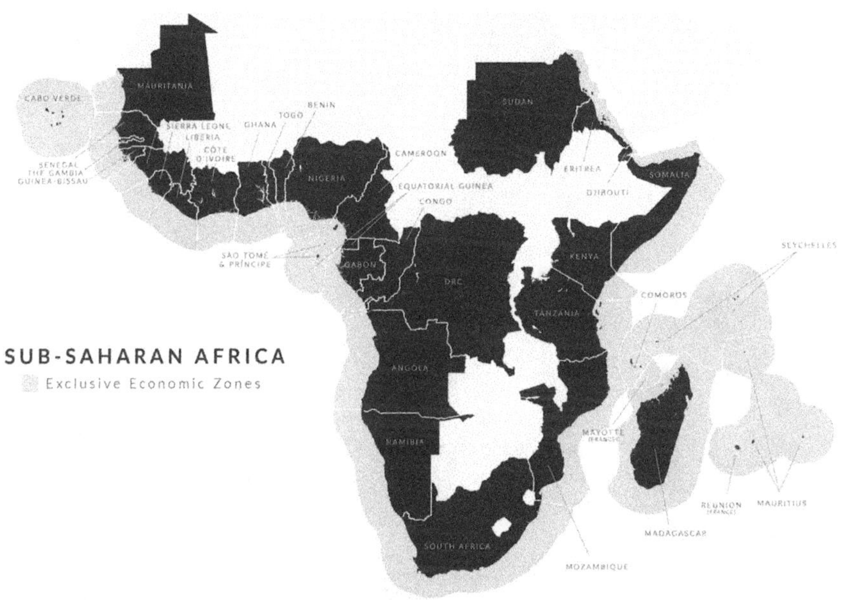

Figure 0.1 Map of sub-Saharan Africa

Introduction

The blue economy of sub-Saharan Africa

Donald L. Sparks, PhD

Until recently, few countries paid much attention to protecting their natural resources. With the publication of Rachel Carson's *Silent Spring* [1] in 1962 more and more people became aware of the fragility of the planet's ecosystem and within the space of a few years a 'green economy' movement became almost mainstream. Like land-based resources, marine resources were also thought to be exploitable, virtually unlimited and there was no need to plan for the future. However, by the beginning of the new millennium there was also a growing awareness of the fragility of the seas and Gunter Pauli's book *The Blue Economy* precipitated a further shift.[2] While their goals and techniques may differ, in many ways both the 'green economy' and the 'blue economy' movements have the theme of sustainability in common.

The World Bank defines the blue economy as one that tries 'to promote economic growth, social inclusion, and the preservation or improvement of livelihoods while at the same time ensuring environmental sustainability of the oceans and coastal areas'.[3] The concept came from the 2012 United Nations (UN) Conference on Sustainable Development (known as Rio+20 or the Earth Summit 2012) that concluded that heathy ocean ecosystems maker for healthier economies.[4]

Considerations about the blue economy are now becoming more mainstream. In 2017 the UN Ocean Conference was focused on finding ways to implement UN Sustainable Development Goal (SDG) 14 ('life below water', as noted in the Appendix). That goal appeals to all nations 'to conserve and sustainably use the oceans, seas and marine resources for sustainable development'.[5]

Of course, the discussion of a blue economy did not begin in 2010 but dates back to the 1950s, if not earlier. The full implementation of the UN Convention on the Law of the Sea (UNCLOS) and related instruments in 1982 was an important step towards ensuring the sustainable development of the world's oceans. As this volume will suggest, the blue economy concept is still evolving.

Interest in Africa's[6] blue economy from academics, government, international organizations and business has increased in the past few years. In 2018 Kenya hosted the first global Sustainable Blue Economy Conference which illustrated the growing importance worldwide of optimizing coastal and marine resources. The blue economy movement was strengthened by the African Union (AU)'s adoption of the 2050 Africa's Integrated Maritime Strategy (2050 AIMS Strategy).[7] Indeed, the AU has correctly called the blue economy the 'new frontier of the African renaissance'.[8]

Thirty-two nations in sub-Saharan Africa have coasts and some 90% of the region's trade is conducted by sea. Maritime zones under African jurisdiction total over 13 million sq km with vast potential for offshore and deep-sea mineral exploitation. Clearly, coastal and marine resources will play key roles as a source of food, energy and economic development in the foreseeable future.

This volume examines a range of inter-related issues divided into three parts. Part 1 introduces the elements of the blue economy in Africa, including cultural aspects, tourism and biodiversity, fisheries, energy and mineral resources, maritime safety, and ports and shipping. Part II presents case studies from three diverse countries, and Part III looks at ways of financing, measuring and governing the blue economy.

We begin with an examination of the region's culture. Due to the geographic and nautical diversity of the region with its distinct east and west coasts, sub-Saharan Africa has a complex and vast maritime history and culture. In addition, the patterns of interaction between people and the ocean and external influences differ, while coastal communities have developed diverse socio-economic characteristics. As history is central to understanding the persistent nature of culture and how it has evolved, these complexities place inherent restrictions on understanding the maritime culture of the region. In order to make sense of the development of distinctive maritime cultural attributes over time, a thematic approach was used focusing on aspects inherent to a maritime culture such as belief systems, the economic use and contribution of the sea, navigation and shipping, maritime power and other interwoven cultural aspects of relevance.

This discussion is followed by an examination of the various ecosystems and the role of tourism. Sub-Saharan Africa supports extremely diverse ecological communities across its eight terrestrial biomes (which include forests, savannahs, woodlands, grasslands, scrublands, deserts and mangroves) as well as multiple freshwater and marine ecosystems. The region's complex climate, geology and history have all contributed to its exceptional biodiversity. Conservation in Africa has gone through major changes over the past few centuries including traditional relationships with nature; exploitation of wildlife and natural resources by European settlers in the 17th and 18th centuries; Western practices of setting aside land shielded from human influences; and more recently integrated conservation and development practices. Africa's conservation biologists and the broader public have shown tremendous fortitude and initiative to overcome the various challenges facing biodiversity over the past few decades. This includes greatly expanding the protected areas network, passing laws protecting the environment and establishing productive partnerships. Historical legacies, poverty, greed, weak governance, competing interests and the consumptive needs of an increasing human population remain the most significant challenges to biodiversity conservation and tourism in Africa. Many of these challenges have led to threats to the future persistence of many species and ecosystems, including environmental degradation and over-harvesting of biological resources. The blue economy has the potential to contribute to make a significant economic contribution to Africa's sustainable future growth and development. However, this cannot be realized unless African governments continue to address maritime security threats and invest in good governance. African tourism comprises only 3% of total world tourism and with 20% of the world's area being sub-Saharan Africa, the value of its tourist trade should have the potential to reach approximately US $400 billion.

Next, we examine the important role of fisheries. African coastal communities have a long history of reliance on fishing and fish farming for subsistence and economic development. Ensuring food security for growing populations is vital, as is sustaining livelihoods, reducing poverty and identifying development opportunities. Many African states have built strong fishery and aquaculture sectors, and have simultaneously developed governance regimes to manage them. More recently, these states have embraced the blue economy to derive further wealth from the oceans, including through expanded fishing and aquaculture activities. While there are clear opportunities, there are also risks associated with placing further pressure on already stressed marine resources and the ocean environment. This highlights the need for enhanced governance to ensure a balance between economic, environmental and sociocultural goals.

Chapter 3 is followed by a discussion on energy and minerals. In the space of less than two decades Africa's hydrocarbon industry witnessed dramatic changes. Between 2000 and 2016 production increased as prices rose. However, from 2017 onwards profits derived from the exploitation of hydrocarbons were lost as prices plummeted. By 2019 the stage appeared to be set for a new wave of dynamism among African nations' policymakers and business communities with falling costs of key renewable technologies creating new avenues for innovation and growth. Chapter 4 discusses the potential hydrocarbon and other marine mineral resources along the continental margin of sub-Saharan Africa. It also includes commentary on non-traditional renewable energy resources, such as natural gas, biomass, wind, tidal and wave.

The seascape surrounding sub-Saharan Africa provides many opportunities for a rich blue economy to thrive. However, maritime-based criminals, such as pirates, human and drug traffickers, and illegal fishers, among others, threaten maritime security and thus the achievement of regional blue economy goals. Preventing criminal threats in the maritime domain is therefore critical. As such, Chapter 5 demonstrates the link between securing the maritime domain in line with the blue economy agenda. The body of this chapter tackles this aim in three sections: first, it reviews maritime crimes common to the region surrounding sub-Saharan Africa that threaten blue economy goals. Second, it critically explores global and regional responses to these regional maritime crimes aligned to the blue economy agenda. Finally, it discusses corruption as a key factor that inhibits the effectiveness of responses to regional maritime security.

Before turning to the country case studies, we examine shipping and ports. The nations of sub-Saharan Africa rely heavily on ports, ships and rail and road transportation. By late 2019 global seaborne trade had gathered momentum and raised expectations within the shipping industry. Global containerized trade increased because dry bulk cargo was in demand, whereas growth in crude oil shipments decelerated. However, during the first quarter of 2020 the beneficial prospects for seaborne trade had dimmed. Risks such as increased inflexible policies and the rise of trade protectionism and the after-effects of the coronavirus (COVID-19) pandemic on the economy are drawbacks to a brighter outlook. Chapter 6 offers an analysis of the region's maritime transportation and examines the challenges that may help or hinder the sustainable development of the infrastructure needed to take advantage of the benefits the benefits that could accrue from the blue economy.

After examining the elements of the region's blue economy, we present three case studies from selected countries. They show a contrast in how different governments have devised strategies for making better use of and protecting their blue economy sectors. They range from South Africa, the region's strongest and most diversified country, to the small island nation of Seychelles. Then Tanzania is examined by way of a contrast to their strategies.

South Africa has a large sea area, abundant marine resources, considerable maritime infrastructure, and ocean-related economic activities contribute substantially to gross domestic product. The growth and development of this sector is an important focus area for the South African government and forms part of an ambitious oceans economy programme known as Operation Phakisa. Chapter 7 is concerned with the attributes of South Africa's oceans economy and its potential contribution to socio-economic growth, development and transformation, as well as the crucial relevance of good oceans governance, security and environmental sustainability.

Seychelles played a leading role in promoting a blue economy concept as a central theme of the sustainable development for Small Island Development States (SIDS) at the 2014 Third Conference on SIDS and has since advocated the blue economy concept in international and regional fora. Chapter 8 presents an overview of Seychelles' experience of implementing a sustainable and inclusive blue economy agenda including the development of the Blue Economy Strategic Policy Framework and Roadmap (2018–30), its early achievements and challenges. Although not a formal evaluation, the chapter draws insights that could inform a global blue economy dialogue and a community of practice.

Tanzania's perspectives on managing the blue economy focuses on the fisheries, marine infrastructure, tourism, energy and trade sectors. The case study notes the abundant but unexploited resources in the Indian Ocean, which include fish, oil and gas, and details the limiting factors which prevent the country from fully harnessing the resources in the ocean for its blue economy. There are many opportunities for Tanzania to develop and prosper from the Indian Ocean through the framework of the AU's blue economy agenda encapsulated in its 2050 AIMS Strategy to enhance a sustainable blue economy for improving Africans' well-being while significantly reducing marine environmental risks as well as ecological and biodiversity deficiencies. The case study draws on information from the Ministry of Fisheries and Livestock, the Ministry of Infrastructure, the Ministry of Natural Resources and Tourism, the Ministry of Energy and the Ministry of Industries and Trade. It highlights areas for better managing and utilizing the ocean resources for poverty reduction and improving Africans' livelihoods. This chapter concludes that managing the blue economy is a function of a proper enabling environment together with political will. In the context of Tanzania, there is political will to engage with the blue economy, but the country lacks a comprehensive and inter-sectoral coordination enabling environment. Thus, the government, in partnership with the Indian Ocean Rim Association member states, should make concerted efforts to support the development of national infrastructure, policy and regulations for harnessing marine resources, research and innovations, as well as the creation of network forums for knowledge sharing in the region.

Part III of the book looks at ways of financing, measuring and governing the blue economy. Innovative finance mechanisms can help to build a sustainable blue economy for sub-Saharan Africa. The ocean economy in a business-as-usual scenario will struggle as a consequence of the ongoing detrimental changes to ocean ecosystems. The coastal countries of sub-Saharan Africa will be particularly affected by declines in their blue natural capital but have much to gain from the blue economy approach. Sustainable finance offers a wide range of concepts, products and approaches to support the necessary transition. Traditional funding from taxpayers will be insufficient to achieve this at the required speed and scale. New finance formats can deliver money faster, at lower risk, engage additional partners, and in particular private sector technology can help to support comprehensive solutions. This cooperative approach is key to addressing the environmental, social and governance challenges that come with delivering an impactful and just transition. Delivered in this way innovative finance can be an important enabler of sub-Saharan Africa's sustainable blue economy.

Following the discussion of innovative financing we look at ways of measuring the blue economy. The availability of consistently estimated data on ocean-related economic values is fundamental to any policy effort to establish a blue economy. Indeed, increased interest in the blue economy concept has been accompanied by an evolving desire to develop regular measures of the contribution that the oceans make to national economies. These measures are being extended to environmental aspects of the ocean economy that will generate a detailed picture of sustainability trends. Adapting standard government data such as the national income and product accounts can provide a basic measurement of ocean-related contributions. Experience elsewhere indicates that this can be done through disaggregation or aggregation measures. Decisions are required about which industries to include and which geographies to cover. Extending this to environmental and ecosystems' aspects of natural capital is important but difficult. Recent development of standardized accounting frameworks, especially those specifically designed for ocean accounting, provide guidance for the creation of such accounts. Despite the difficulties and resource requirements, it is critical to start with what is available as soon as possible.

The emergence of the blue economy as an aspirational development framework for sub-Saharan Africa has taken place in conjunction with its rise on the international stage. Understanding the international ocean governance framework underpinning the sustainable blue economy concept contextualizes these ambitions when balancing inherent tensions between growth and sustainability perspectives. In Chapter 12, key multilateral agreements relevant to the sustainable economic development of maritime space and marine resources, both within Africa and globally, are summarized with a particular focus on the environmental components of these agreements. The national implementation of international and regional agreements, together with institutional strengthening, policy harmonization and cooperative approaches between states, will help to achieve the full potential of a pan-African sustainable blue economy. Experience and commentaries to date suggest that for blue economic growth to be truly sustainable, its governance must be situated within ecological and social realities, over a range of spatial scales and time frames, with progress regularly assessed and adjustments made. Socially equitable approaches underpinned by scientific knowledge and embedded within an adaptive and holistic governance framework will have the highest chance of lasting success.

The final chapter in this volume looks at making the blue economy possible. Sub-Saharan Africa's blue economy policies are influenced by global instruments such as UNCLOS, as well as global targets as set out under the SDGs. While many sub-Saharan African countries have signed up to UNCLOS, very few have taken the next step of domesticating the framework law into their national policies. Furthermore, few countries are making use of tools such as Marine Spatial Planning (MSP) and Ecosystems Based Management (EBM) to inform their policymaking and to advance their blue economic activities. A number of island countries such as Mauritius and Seychelles have significantly extended their maritime boundaries based on some provisions of UNCLOS, while others such as South Africa have ambitious blue economy programmes through Operation Phakisa. The majority of sub-Saharan African countries are not engaged in the exploration or exploitation of resources in their exclusive economic zones (EEZ) or in the community heritage areas, despite having enabling global policies and strategies such as UNCLOS and the 2050 AIMS Strategy, as well as planning tools such as MSP and EBM at their disposal. It is possible to significantly grow sub-Saharan Africa's blue economy if countries fully embrace and domesticate global laws and agreements that give them access to resources in the common heritage areas and in their EEZ.

Finally, as an afterword, we turn to what is possible. Of course, as will be shown, what is possible and what is likely are not necessarily the same, especially in the short term given the consequences of the COVID-19 pandemic.

Notes

1 Rachel Carson. *Silent Spring*. New York: Houghton and Mifflin, 1962. It should be noted that Carson's earlier works were about the oceans. They included *Under the Sea-Wind* (1941), *The Sea Around Us* (1951) and *The Edge of the Sea* (1955).
2 Gunter Pauli. *The Blue Economy*. London: Paradigm Press, 2010, and *The Blue Economy 3.0*. New York: Xlibris, 2017.
3 *The Potential of the Blue Economy*. Washington, DC: World Bank, 2017, p. vi.
4 *Blue Economy Concept Paper*. New York: United Nations Department of Economics and Social Affairs, 2014.
5 https://sdgs.un.org/goals/goal14.
6 For the purposes of this study, sub-Saharan Africa comprises the 53 states located south of the Sahara Desert and the term 'Africa' is used as shorthand.
7 *2050 Africa's Integrated Maritime Strategy*. Addis Ababa: African Union, 2012.
8 *For the Launch of the 2015–2025 Decade of African Seas and Oceans and the Celebration of the African Day of the Seas and Oceans on 25 July 2015*. Addis Ababa: African Union, 2015.

Part I

Elements of the blue economy

1 Culture, communities and society

Thean Potgieter, DPhil

Introduction

Africa's history and future are intertwined with the sea: '*Bad nin tegey yaab, waxuu ku warramana ma yaqaan*'.[1] The sea is an inherent part of its culture, economy and existence, historically connecting the continent to the rest of the world. Contemporary sub-Saharan Africa is the product of its pre-colonial, colonial and post-colonial experiences. Maritime interaction, from ancient contact with Arabian and Eastern civilizations to mercantile Europe, had a great influence on the region's history. As mercantile and political interest continued to grow and expand, what started as coastal partnerships and a thirst for gold, ivory and slaves evolved into colonialism and the current quest for raw materials. After a complex narrative of development, resistance and subjection the second half of the twentieth century introduced an era of nationalism, triumph, freedom and realignment. Africa had to re-establish its relationship with the world and within the context of its capacity and limitations find new horizons. Crucial determinants through the centuries were Africa's access to the sea and its own wealth, which made it both receptive and vulnerable to what the sea brought.

Interaction with the oceans is more limited in the history of sub-Saharan Africa than in that of the Mediterranean and certain parts of Asia, for example. Reasons for this vary, but it may be partly explained by the abundance of much of Africa's soils, its relatively small populations, and vast distances from potential markets, compared to the concentred coastal populations and easy interaction across small stretches of ocean in other parts of the world. In addition, there are few natural harbours (the United Kingdom has more than the entire region) and navigable rivers leading to the sea. The recent history of sub-Saharan Africa is inward-looking and reflects 'sea blindness'. This could to some extent be ascribed to the multiple challenges that the continent's newly emerging states faced in the post-colonial era. However, the region needs to focus more closely on the sea and the inherent wealth of its maritime resources.

It is evident from the introduction to this chapter that sub-Saharan Africa has a complex maritime culture. But what is meant by maritime culture?

'Maritime' implies human activity related to the sea; it is connected with ships, the sea and being near the sea. 'Culture' denotes a way of life, specifically the general customs and beliefs shared by a particular group of people, evident in behaviour, habits, attitudes, language practices, morals and religious beliefs.[2] Cultural meaning evolves from shared experiences over time and permeates societal activities. It is not just a set of rules, or an interpretation of symbolic and ritual acts, but it is present in every field of human endeavour from politics and economy, to kinship and beliefs. As the interaction between society and its past is unending, history contributes significantly towards 'shaping culture and its evolution'.[3]

In this discussion the maritime culture of sub-Saharan Africa is understood as the heritage, legacy and traditions inherited from past generations, focusing on human interaction with the ocean. As shared historical experiences influence cultural practices, a sense of historical processes and the evolution of societies is crucial. The chapter argues that the sea is an inextricable part of sub-Saharan African culture, represented in going to sea, coming from across the sea, belief systems, the economic exploitation of the sea, artefacts, maritime power, and shore-based practices that have evolved due to human interaction with the oceans.

Community, spirituality and the sea

Africa's diverse spiritual and religious traditions have a central place in all spheres of life, social, economic and political, and it is said that African people are 'notoriously religious'.[4] The three main religious heritages – African traditional religion, Christianity and Islam – have a long history and exert considerable influence on the continent and are all associated with the maritime environment.

Traditional African belief systems can be traced back to the emergence of African societies. Humanity is associated with lineage and community, which define people's spiritual origins and create uniqueness informed by ethnic identity. Membership of a community, and its associated language, beliefs, habits and memories, shapes consciousness and is not necessarily a political unity. The unifying concept is kinship, as the source of wisdom is attainable through ancestors, and specific lineages link living people to the spirits of those that went before.[5] Spirituality is a core African value, and it is not to be understood in the Western, Eastern, Christian or Islamic conception of a deity or gods.

One important aspect of this spirituality revolves around water: it is recognized as being essential to life and is a source of strong spiritual power. Water spirits are linked to rivers, wetlands and the sea and these have imparted gifts linked to healing, sacred knowledge, psychic abilities and medicinal plants to healers and diviners who are the custodians of traditional knowledge.[6] Water deities in African tradition belief systems are linked to fresh water as well as to the sea.

Although various examples exist, the following instances could briefly be highlighted. Beliefs about water spirits, particularly those related to the snake and the mermaid, and their role in the calling of traditional healers, are complex in Southern Africa. The dwelling places where these spirits reside (deep pools, often below waterfalls, fast-moving water, or the sea) are sacred and are key to the training of healers and the performance of family rituals.[7]

The age-old religious customs of the Yorùbá people of Western Africa are a wide blend of indigenous beliefs, myths, legends, proverbs and songs influenced by the cultural and social contexts of their region.[8] The sea is a constant source of reflection and explanation about the world and the unknown. In the Yorùbá religion, the ancient meaning of Olokun (also referred to as a sea god) is vast; she is the owner of great waters and a symbol of the ocean depths and seas. Olukun is linked with the Mother of Fishes, Yemoja or Yemonja, and they represent abundance, fertility and prosperity, as well as the sources of life and mystery.[9] Yorùbá practices were taken across the Atlantic by enslaved Africans, influencing belief systems as part of the African diaspora. In Santeria (an Afro-Caribbean religion based on Yorùbá beliefs), Yemaya is the Ocean Mother Goddess and is often depicted as a queenly mermaid. Anything from the sea is a symbol of Yemaya as her energy comes from sea rocks and shells.[10]

Despite colonialism and the introduction of Christianity and Islam, traditional belief systems remain prominent as some coastal fishing communities in Ghana still embrace traditional beliefs related to the sea and fishing.[11] Of note is the belief that water symbolizes peace, fertility and growth and that the sea is a deity protecting and overseeing fishing. The sea is a custodian of morality and bad luck can be ascribed to punishment by sea deities for violating fishing taboos.[12]

Christianity and Islam were both partly introduced through maritime interaction. The early introduction of Christianity to North Africa, apart from Ethiopia, was followed centuries later by its reintroduction as a result of European voyages round the continent. From the seventh century onwards, Islam spread by way of Indian Ocean trade routes and across the Sahel. Although trans-Saharan trade between West Africa and the Mediterranean predated Islam, it was strengthened by North African Muslims and Islam gradually spread to coastal states, including present-day Senegal, The Gambia, Guinea and Nigeria. It went along with the written word: through Arabic, the language of the Quran, West African Muslims documented events and culture, while Islamic writing and calligraphy inspired art. As a result of colonialism Arabic lost some of its political and administrative importance.

As Swahili settlements had much contact with each other and the Arabic world, together with language, the influence of Islam spread extensively and created a common bond between those already associated with the ocean, with worldviews matching customs and practices. Seafarers and traders were at ease with one another because, despite their political differences, they shared commonalities in language and faith.[13] Although East Africa does not have centres of learning and Islamic culture as does West Africa, a notable medieval traveller, Ibn Battuta, describes fourteenth-century Mogadishu as a prosperous trade port known for its mosque, which inhabitants of the region visited specially on certain dates.[14] Mosques are ubiquitous to Swahili settlements. Many have been in use for centuries or are now famous as ruins such as those of the Great Mosque of Kilwa in today's Tanzania.

Roman Catholic missionaries followed the first Portuguese traders along the coast of Africa from the fifteenth century onwards, which contributed to the spread of the Portuguese language, culture and influence. As European engagement with Africa expanded, missionaries from other countries and Protestantism followed. Christianity also contributed strongly to the anti-slavery stance.

Trade and commerce

Trade and commerce are ingrained elements of the African maritime cultural landscape. Pre-colonial traditions included the Sahel trade as well as the long history of seaborne trade along the east coast of Africa and the associated contact with the Omani Arabic world, India and China. Following the first Portuguese voyages along the west coast of Africa, commercial contact with Europe and the world increased. In maritime terms, though, the history of the south coast of Africa is different. Due to its geographic location and sea conditions the region was not valued for coastal trade; however, its location between the Indian and Atlantic Oceans, astride the long sea route that carried enormous riches from the East, made it an ideal strategic refreshment post for shipping.

Trade and commerce: the east coast

Reference to the African Indian Ocean trade exists in ancient texts. The *Periplus Maris Erythraei* ('Voyage around the Erythraean Sea'), an anonymous first-century CE Egyptian-Greek text offers some first-hand historical perspectives of the east coast of Africa as far south as the coast of present-day Tanzania.[15] The *Periplus* described coastal trade as well as trade relations between East Africa and the Greek-Roman world (through Egypt), Arabia and India. Helped by the predictable monsoon winds, Arab traders were established along the coast and commodities from Africa that were in high demand included ivory, myrrh, rhino horn, palm oil and slaves.[16] Later Arab descriptions point to Arab trading posts along the east coast of Africa growing in size, wealth and importance, while trade in gold, slaves and ivory, among others, took place with the Arabian Peninsula, Oman, the Gulf region, the Ottoman Empire, India and China.[17]

Opinions about the origins of the Swahili differ. Is it only the maritime culture that developed in the first millennium CE or is it inclusive of the earliest known Stone Age settlements along the coast? Although these questions are relevant, of note is that an age-old African maritime culture developed across the socio-economic spectrum. It is an important example of how maritime interaction influenced the African cultural landscape.[18] The Swahili population was Muslim but not Arab, and the culture could be seen as a fusion of African society with Arabs and other lesser influences. As trade was the lifeblood of the Swahili settlements they grew in strength, shared ideas and customs, specifically in the cultural, linguistic and religious spheres. By the fourteenth century the many Swahili trade ports along the coast included Mogadishu in the north, and Lamu, Malindi, Mombasa, Zanzibar, Kilwa, Mozambique Island and Sofala in the south.[19] These vibrant trading ports were like 'stepping stones along the Indian Ocean'.[20]

African imports from Asia included cotton products from India, Malay spices and ceramics from Persia and China,[21] while archaeological traces date a Chinese presence along the coast back to the eighth century. Early Chinese sources list major traded goods from Africa as gold, amber, yellow sandalwood, ivory and slaves, while even a giraffe made its way to the imperial court in China. Between 1405 and 1433 large Chinese fleets made several voyages to the east coast of Africa. Although these voyages were interrupted by domestic events in China, Arabian, Persian and Indian intermediaries ensured that Chinese products such as porcelain and silk were still available on the East African market.[22] After the Portuguese arrived on the scene at the end of the fifteenth century, they dominated parts of the coast and its fortunes for about two centuries.

The history of the Swahili is of particular relevance as the example of Kilwa highlights. The ruins of the mosque are evidence of commercial prowess and a powerful city-state in the first half of the second millennium CE. Kilwa was central to the trade of the most important commodity of the region – gold.[23] Gold was imported via Angoche and Sofala from the mines in present-day Zimbabwe and was traded with merchants from Arabia, Persia and India.[24] Warehouses were built for the import and export trade and by the thirteenth century Kilwa had begun to strike its own coins.[25] Early Portuguese visitors recorded the weighing of gold and that duties had to be paid at Kilwa.[26]

After Saīd ibn Sulṭān (also referred to as Seyyid Said) became the ruler of Oman in 1806, he expanded his economic and political influence across East Africa and introduced an era of trade growth and renewed prosperity. In 1840 he moved his capital to Zanzibar, and established large clove plantations which relied extensively on slave labour. Indeed, at that time Zanzibar was the world's largest producer of cloves. Zanzibar exported to Arabia, India, the Far East, America and Europe, among others, and imported manufactured goods to trade along the east coast and within the African interior.[27]

Maritime trade stimulated interaction between coastal regions and the interior. Elaborate trade representation and routes were established, and the

Swahili culture reached inland to the Great Lakes region. Goods, such as ivory, gold, copper, ebony and pottery, were brought from the African interior to ports and trading posts. Merchants from Zanzibar cooperated with Swahili and other traders from the mainland and from the seventeenth century onwards Oman became more dominant in the slave trade, which in Central and East Africa was essentially in Arab hands. In the nineteenth century Zanzibar became the main centre of this 'ruthless, flourishing and well-organized business', which was bigger than ever.[28]

Due to the high demand for ivory and slaves, traders penetrated Central Africa, the Republic of the Congo and even Angola. Tippu Tib was a typical late nineteenth-century powerful and ruthless slave and ivory trader. With a private army he controlled a vast area stretching from Lake Tanganyika and the Ituri forest into the Congo Basin as far as Basoko. He facilitated the colonization of the eastern part of Central Africa and was even appointed governor of the Falls region in the Congo Free State by the Belgians.[29] However, he posed a threat to the colonizers, and the Belgians defeated his forces in 1893. Tippu Tib withdrew to Zanzibar and wrote his autobiography in Swahili.[30]

The actual extent of the slave trade is still debated. It is estimated that by 1810 roughly 10,000 slaves were sold at Kilwa and Zanzibar annually, and that by the 1860s between 30,000 and 35,000 slaves arrived in ports controlled by Zanzibar,[31] whereas some estimations indicate that the 1860s 'figure had risen to 70,000 for Zanzibar alone'.[32] By 1870 Oman imported 13,000 slaves per year, transporting them to the Persian Gulf and Persia, Mesopotamia or Baluchistan and India.[33] Visitors to Zanzibar remarked on the 'brutality' of Arab masters towards African slaves and this trade is noted for being 'most tragic and cruel': African women often became concubines, and some sources indicate that the castration of male slaves, thereby rendering them eunuchs, 'was common practice', with many bleeding to death in the process.[34]

Due to European involvement, colonization and the 'scramble for Africa' during the late nineteenth century the character of trade soon changed with the occupied territories becoming marginal participants in expanding colonial economies.[35] In the decades that followed the Second World War, countries in the region gained their independence. Although the colonial era was but a short period in the history of Africa, it made a huge impact. The powerful Swahili trading states are long gone, and trade between the coastal countries of East Africa is not on par with its historic precedent.

The maritime legacy remains evident in Swahili culture, in its way of life, socio-economic and political activities, language and religion. The interface between humans and the sea, commerce and immigrants, caused initial contrasts to diminish and over time fused a unique African maritime culture.[36] Even early visitors reflected on separate Swahili states, either rivals or allies, that were linked by a common historical course, trade, sailing ships, monsoons, language and religion.

Trade and commerce: the west coast

The Sahara Desert was a thriving crossroads of exchange and trade between West Africa, North Africa, the Middle East and the Mediterranean. After Islamic conversions in West Africa during the seventh and eight centuries, trans-Saharan camel caravan trade routes grew and peaked during the sixteenth and the early seventeenth centuries. During the gruelling journey south camel caravans carried glass vessels, beads, glazed ceramics, copper, books and even salt. Trade was fuelled by gold from West Africa, famed for its pure quality and used for minting coins and adorning luxury items and religious objects. The ruler of the Mali Empire, Mansa Musa, was renowned for his wealth and as a principal supplier of gold to the Mediterranean world.[37] Besides gold, the caravans heading north transported ivory, animal hides, leatherwork, spices and captives forced into slavery.

The fabled sub-Saharan African gold, the promise of spices and the silk trade with Asia drove Portuguese seafarers to explore the African coastline from the fifteenth century onwards. This introduced new trade possibilities and heralded the Atlantic era in world history. As the Portuguese focused on trade, not colonization, they established strategically placed coastal trading forts or 'factories' along the coast, exchanging textiles, iron and weapons, for gold, pepper, myrrh and slaves.[38] The initial Portuguese objective was gold, but with the creation of sugar plantations in Brazil and the Caribbean the slave trade became dominant. The Portuguese established the fort of São Jorge da Mina (Elmina Castle) in 1482, which became a symbol of Portuguese power and an important holding place for slaves before their transportation to the Americas.[39] As Portuguese influence was essentially limited to their coastal forts, trade was conducted through agents. Much of the extensive trade system that the Portuguese encountered lay in the hands of Muslim merchant groups generally known as Juula who traded as far as the tropical rainforests.[40]

Although a greater understanding of the economic history of pre-colonial sub-Sahara Africa and the presence of its many capitalist elements is still required, the use of coins and currencies provides interesting evidence about trans-Saharan trade, coastal trade and interaction with the interior. The chronicler of Cordova, al-Bakri, reported in 1067 that the tax system of the king of ancient Ghana was based on regular units of value, whereas graded weights of gold dust and Indian Ocean cowrie were used as standard measures.[41]

Unlike East Africa, West Africa was not as socially homogeneous or as economically unified, and was characterized by social, linguistic and religious divides, while significant differences in local economies and subsistence production existed. Nevertheless, interlocking zones of trade and currency circulation did exist which provided a complex but commercially rich environment suitable for sustainable trade between Africans and Europeans.

A century after the arrival of the Portuguese, other Europeans followed. It reshaped the structure of trade and increased the commercial dynamics between the coastal regions and the interior. Commercial practices expanded, creating new networks, trade patterns and measurement calculations. European traders had to adapt to local trade complexities, practices and values, and focus on the goods that African traders wanted. As the social protocols and courtesies of commerce had to be adhered to, visitors had to demonstrate proper respect to African trading partners and be well behaved and trustworthy.[42]

Forts were not exclusive European enclaves, as Europeans, Euro-Africans and Africans visited, lived, traded and worked in these evolving settlements. European merchant ships and local trading vessels linking coastal towns and islands came and went by sea.[43] Although a truly global trade network involving Africa, Europe and the Americas came into being, it was also a brutal trade based on the individual tragedies of millions of captives forced onto ships, destined to be slaves in the Americas.

When the first Europeans visited the Upper Guinea Coast and the Bight of Benin in the Gulf of Guinea, an export trade in locally produced cotton items was sustained by cotton production. The origin of the weaving industry is not completely known, but by the eleventh century al-Bakri could describe cotton textile manufacture in the region. It was well developed by the fourteenth century and by the early sixteenth century European traders held Yoruba hand-woven cloth and the quality of its dye in high regard, buying it to sell in Europe. Cloth weaving evolved due to demand, but it became an important economic resource and a currency with purchasing power. In the Yoruba culture it was also a means of adornment and a part of art.[44] Although the increasing demand for European fabrics later killed the industry, it enjoyed a post-colonial resurgence due to pride in Yoruba ceremonial clothing.

However, it is the Atlantic slave trade that lives on in infamy. Spanning the period between the sixteenth and nineteenth centuries it is historically the largest long-distance forced movement of people. It was initially controlled by the Portuguese, but from the early seventeenth century onwards other European countries also became involved and an array of fortified trading posts dotted the coast from Gorée Island (present-day Senegal) to the Niger Delta. As the Portuguese began to be ousted, they traded further south towards the Congo Basin and Angola. Sources differ as to the precise numbers associated with this terrible trade, but it is estimated that around 12.5 million enslaved Africans were transported across the Atlantic with about 10.7 million surviving the notorious ocean voyage. Ships carried the flags of many nations, but the greatest volume of enslaved Africans was transported in Portuguese, British and French ships.[45] As most slaves were sold by Africans from Central and West Africa to European slave traders, the coastal communities played a significant role. The loss of population in Africa was not evenly shared as some areas suffered disproportionally, and others profited from the trade, yet it destroyed people and whole cultures, destabilizing and changing the African continent forever.[46] Evidence suggests that the total number of persons deported as slaves from Africa between the eighth and the mid-nineteenth centuries reached at least 24 million.[47]

Although anti-slavery sentiments became common by the late eighteenth century, it was still defended in certain British circles as being 'beneficial to the nation'.[48] Soon the context of the French Revolution and the Enlightenment gave new impetus to the campaigns to abolish slavery. Britain banned the importation of African slaves to its colonies in 1807, but slavery was only abolished in 1833. By the nineteenth century slavery had been widely abolished in Britain and the United States followed in 1865.[49]

The drastic decline of the Atlantic slave trade implied a crisis to the export economies of West Africa and alternatives had to be sought. Palm oil was one such commodity. Trade boomed and the waters of the Niger Delta, from whence ships had previously embarked with their human cargoes, became known as the Oil Rivers (a reference to the later impact of fossil fuels). The same applied to the Dahomey coast (also referred to as the Slave Coast) and the Gold Coast. Between 1862 and 1872 the trade in palm oil exports and cotton cloth imports between Britain and the Gold Coast quadrupled. As Britain manufactured margarine from palm kernels, annual exports from Lagos rose to roughly 37,000 metric tons at the end of the century. In the Senegalese-French trade, peanuts became a source of oil.[50]

Another boom occurred in cocoa production. The plantations on the Portuguese islands of São Tomé and Príncipe were notorious for the slave-like working conditions of more than 67,000 persons shipped from the continent between 1888 and 1908.[51] Cocoa trees were planted along the Gold Coast and exports jumped from about 36 kg in 1891 to about 179,000 metric tons in 1919. Imports kept pace with exports and in the last half of the nineteenth century the contemporary value of the annual British trade with West Africa (Nigeria, the Gold Coast, Sierra Leone and The Gambia) climbed from about £1 million to more than £9 million per year. The switch from sail to steam ships boosted trade volumes, and British trade with West Africa rose from about 55 tons in 1854 to about 540,000 tons going through Lagos alone by 1900.[52]

Of note is that these essentially pre-colonial economies and societies proved themselves capable of coping with expansion and new demands. Davidson argues that these changes were not a 'revolution in the indigenous economies',[53] but should be ascribed to the fact that the overseas economies trading with the region moved beyond the slave trade and provided African economies with the opportunity to develop their inherent growth potential. During the colonial period these economies were controlled by European powers. After independence it was extremely difficult for the new African countries to build multifaceted indigenous economies and to break the pattern of providing raw materials and being major consumers of manufactured products. Essentially, many remain 'periphery' economies, providing raw materials to the 'centre'.[54]

Power, politics and conflict

Due to the wealth associated with the commercial use of the sea and the importance of maritime strategic locations, power, politics and conflict are inherent to the complex interwoven whole that makes up the multifaceted maritime history of sub-Saharan Africa. The sea brought conquest and colonialism. The region regained its independence, but as the maritime interests of African states are often threatened political, strategic and military aspects remain central.

Since time immemorial, piracy and theft have been associated with the rich cargoes that ships carried across the oceans, or in ports and anchorages. Sub-Saharan Africa is no different as its extensive coastline and adjacent islands provided ample hiding places for pirates frequenting the coast from the Red Sea to the western extremities of the continent. The *Periplus* refers to ships being attacked by 'men of piratical habits', while sources throughout the centuries abound with stories of piracy and raiding.[55] Piracy became more widespread due to the rich European trade with the East, while some pirates became slave traders and even participated in local wars.[56] In the wars between the European maritime powers and in the Napoleonic era Indian Ocean privateers preyed upon trade as far south as the Cape of Good Hope.

In the early twenty-first century the scourge of piracy is not uncommon in historical hotspot areas and the increase in the number of acts of piracy and kidnappings off the west coast of Africa are of concern.[57] Somali-based piracy has sparked international anxiety, but the experience has reinforced old lessons: it is crucial to fight piracy and maritime security at sea, while also restoring law and order on shore.[58] In this case international and regional cooperation are crucial in limiting Somali piracy.

After the Portuguese arrived on the east coast, trade competition and conflict erupted and Swahili power declined. Within 12 years the Portuguese had shattered a civilization, sacked cities and established their sovereignty over part of the coast.[59] The technology and military systems of the Renaissance and early modern Europe gave the Portuguese and European powers a significant military advantage. Nothing at sea could match their large sailing ships armed with heavy guns, and ashore European steel weapons, iron protection and superior firepower added to the military power of soldiers and the strength of coastal fortifications.

In 1593 the Portuguese started to build a huge fortress called Fort Jesus in Mombasa. As their authority was always contested by the local Swahili and Arabs, during the seventeenth century they lost the fortress to the local ruler, only for it to be captured by the Sultanate of Oman in 1698. The Portuguese briefly retook the fortress in 1728, but lost it a year later to Omani forces. Coastal affairs would now be dominated by the local Swahili and the Arabs from Oman. After the dynamic Seyyid Said became the Sultan of Oman, trade and power relations in the western Indian Ocean shifted again. By 1837 the coastline, from Cape Delgado in the south to Pate and Lamu in the north, was under Omani control. In 1840 Said moved his seat to Zanzibar.[60]

When the politics of slavery changed, the British Royal Navy and other navies contributed greatly towards restricting the trade. In the Indian Ocean persistent British efforts to crush the Arab slave trade with naval patrols, diplomacy and blockades led to its decline. The Royal Navy ceaselessly patrolled the coast and by the late 1890s some of the last vestiges of the East African slave trade had been destroyed.[61] However, a new era arrived as much of the east coast from Kenya to South Africa, including Zanzibar and many of the Indian Ocean islands, were now under British, Portuguese, French or German control.

Many of the early Portuguese expeditions along the west coast were military in nature, with the Portuguese raiding local populations for goods and slaves. After the initial surprise, the Portuguese began to meet organized resistance, often from boats carrying 70–80 armed men. Portuguese sources mention ingenious ambushes and determined defence causing them to withdraw without fighting. The overriding goal was not war, as the Portuguese wanted to catch slaves, but abducting people had now become too dangerous.[62] The result was the development of complex trade networks, with African polities often using force to obtain captives for the slave trade.

Following the abolishment of the slave trade, the British Royal Navy joined forces with other navies to really bring it to a standstill by policing the west coast of Africa and seizing slave ships. The British also concluded treaties with African chiefs and destroyed slave trading facilities onshore, using the pretext of fighting the slave trade to establish new trade and extend British influence.[63]

The 'scramble for Africa' was in essence an exercise in maritime power projection. Due to the first two Industrial Revolutions and a widening technological gap, Europe had the means and will to project power across the world which resulted in invasion, occupation, division and colonization. None the less, to brush off military operations as European successes due to superior weaponry does a disservice to the many competent African commanders, their tactical prowess, staunch resistance and military organization. Operations did not always result in victory to the colonial forces as the Anglo-Ashanti wars and battles such as Isandlwana and Spion Kop dispelled the notion of Europe's inherent military superiority. After the Ethiopian Emperor Memelik concluded a boundary treaty recognizing the Italian coastal colony of Eritrea in 1889, the Italians still tried to extend their territory into Ethiopia, but were defeated by an Ethiopian army at the Battle of Adowa in 1896. Yet, by the time that the First World War commenced European colonial powers had carved up Africa (with the exception of Ethiopia and Liberia). Their armies and navies succeeded in crushing the initial resistance to colonialism due to a number of crucial variables including technology, strategic and tactical adaptation, military organization, discipline and logistics.[64]

A culture of defence against maritime power projection developed in sub-Saharan Africa. The 'scramble for Africa' was about the formal conquest of the continent and the chance of it occurring again is slim. Even so, Africa is now defending itself against a 'new scramble for Africa', but this time it is about the continent's resources on land and in the maritime domain. The preparedness of the continent to fully defend its interests this time around is debatable. As powerful economic and political forces, compounded by illicit activities, are now at play, African defence lies in good governance, maritime security and developing a culture of cooperation among African states.

Navigation, ships and boats

Navigation in the Indian Ocean was pioneered by the Arabs. Through constant observation of tides, currents, winds, monsoons and atmospheric conditions, a thorough appreciation of the nautical environment developed and was recorded. Such data made it possible to traverse the Indian Ocean in early boats.[65] Indian Ocean navigation significantly improved between the thirteenth and fifteenth centuries as Chinese knowledge added astro-navigation and the compass came into use (probably also borrowed from the Chinese), while early charts were drawn up with annotations on nautical conditions.[66] Between November and March seafarers exploited the north-eastern monsoon to sail from Arabia and India down the coast of Africa, to return with the south-western monsoon between May and October.[67]

In much of the Indian Ocean weather conditions were moderate and predictable. As beaches along the African coast were mostly sandy, boats were constructed to be beached at high tide and cargoes were disembarked at low tide. Protected bays, coral reefs and offshore islands provided ample shelter for larger vessels to anchor and disembark cargoes into smaller boats.[68]

In terms of navigation aids, emphasis often falls on the lighthouse towers at Barawa and Mogadishu as well as the Mbaraki Pillar at Mombasa, which the Portuguese described in 1593 as a navigation aid for ships entering the port.[69] Furthermore, substantial efforts were made to create a series of artificial coral reefs and sand causeways perpendicular between Kilwa and Lindi, as well as the causeways that lined the approaches to Kilwa where the shores of low-lying islands were heightened with coral.[70] Although still a point of debate, they were most probably created in the early fourteenth or fifteenth centuries to aid navigation for ships on Sofala-Kilwa gold run.[71] Kilwa had sea walls, a safe natural setting and was easy to approach under sail in monsoon conditions. Due to import and export operations and bulk cargo aggregation warehouses and port infrastructure also developed.[72]

A long history of sailing directions or guides in a large variety of languages exists for the whole coast of Africa. After the first European voyages around Africa, nautical data were recorded, hydrographic surveys took place and new charts were published. By the early seventeenth century seafarers were warned of navigation dangers and fully utilized global weather patterns to maximize voyages around Africa and between Europe and Asia.

Simple dugout vessels and wicker or sewn boats were common in sheltered coastal waters and were the most basic traditional boats. Slim-hulled outrigger sailing canoes were also developed as well as boats with raised bows and higher sides that could sail on the open seas. The Austronesian influence in the transfer of maritime technology is also evident in the style of outrigger canoes found along the east coast of Africa, the Comoros and Madagascar.[73]

Although vessels from the Arabian Peninsula initially plied the Indian Ocean trade routes, boat-building skills in the western Indian Ocean soon evolved and as a thirteenth-century Arabic manuscript indicates it is highly likely that the inhabitants of East Africa also plied the same trade routes.[74] Typical vessel designs ranged from small boats with a curved bow and stem, to larger vessels with raked or flat sterns.[75] Over the centuries these vessels were known as 'dhows', a generic term of Arabic origin, as the design can probably be traced back to Oman. The term is associated with fore-and-aft rigged sailing vessels with one or more masts, with lateen sails on long yards, a pointed bow and a square stern. Due to their seaworthiness, the availability of construction materials, easy maintenance and the efficiency of the lateen rig, the construction and use of such vessels became commonplace in the Indian Ocean. No nails were used as planks were sewn together with coir (often from the coconut palm), sails were made from plaited palm leaves and the boats were caulked with pitch.[76]

An exception to the common sail-plan is the square-rigged *mtepe* with a long, curved prow and pointed stern. These vessels were seen along the east coast up until the early twentieth century, and were more dependent on monsoon conditions as sailing close-hauled with squire sails is more demanding. Weighing up to 40 metric tons and measuring 30 metres in length, they undertook offshore voyages to transport bulk cargoes, building poles, rafters and mangrove timber.[77]

Lateen-rigged vessels ranged from small fishing or coastal vessels to ocean-going cargo ships with up to three masts.[78] The Portuguese recorded in 1505 that trading vessels in Kilwa were as large as their 50-metric-ton caravels.[79] As they could sail closer to the wind and were more manoeuvrable, lateen-rigged vessels became more commonplace. Such vessels are still built and used in the coastal waters of Africa, the Arabian Peninsula and the Indian Ocean islands – even making commercial journeys under sail.

Although the pre-colonial West African history of oceanic trade and the development of ocean-going vessels are not as elaborate, fifteenth-century Portuguese sources refer to large boats build by the Bulom and Bijago on the coast of upper Guinea. These boats could carry between 90 and 120 persons, which gave the Bijago (who lived on fortified islands) naval superiority, making it possible to raid the mainland.[80]

Although studies on the construction of traditional boats are important for understanding the maritime culture of sub-Saharan Africa, these are also artefacts relevant to the development of broader society.[81] Much more understanding of the evolving maritime traditions of Africa is required.

Fishing

Although fish is an important source of protein, sustainable fish stocks have decreased globally from 90% in 1974 to 65.8% in 2017.[82] In Africa the ocean provides vital nutrition to more than 200 million people, an average of 22% of their protein, and an income to more than 10 million people.[83]

Sub-Saharan African coastal communities are no different to the rest of the world, as these communities have been engaged in sea fishing throughout history and seafood is an ancient source of protein. The ecological diversity of the African landscape and coast stimulated production: in the hinterland agriculture and hunting were common, while coastal communities relied on fishing and shellfish beds, as well as mangroves and coral for construction.[84]

The *Periplus* refers to fishing and the catching of tortoises from sewn boats and canoes hollowed from single logs, as well as the use of wicker 'nets' which were secured between channel-openings to catch fish. In centuries to come Arab authors and Portuguese seafarers emphasized the importance of seafood for local consumption and trade, as well as the commercial gathering of pearls, shells, turtle shells and amber.[85] Evidence from the east coast of Africa suggests that consumption of open ocean species was considerably rarer compared to fish found closer inshore and around coastal reefs. The old fishing techniques, using lines and hooks, fish traps and nets, are still common today.[86]

Along much of the west coast, specifically in the accessible estuaries and river mouths, fishing was a highly specialized way of life. As early fishermen such as the Fante (who still live in present-day Ghana) were renowned for their skill in handling canoes in heavy surf, they were recruited for loading and offloading cargoes.[87] However, the disruption brought by profits from the slave and ivory trade would have had a negative impact on traditional economic activities.[88]

Oral traditions and research indicate that the fishing folk of West Africa have continued to adopt new methods and means to improve their craft. The switch from sails to outboard engines and the use of larger ring nets in the mid-twentieth century meant that African fishermen and communities flourished. As their range increased, they were able to access larger stocks of fish, and by storing catches in ice boxes they could stay out at sea for three to five days.[89] Sadly, though, in the late twentieth century catastrophe struck, as large industrial trawlers and factory ships engaging in illegal, unreported and unregulated fishing plundered the fishermen's traditional rich fishing grounds.

Marine fishing is 'more than a livelihood; it is a way of life' to many Sub-Saharan coastal communities.[90] Local communities must therefore be partners in fisheries, environmental management and maritime security, while fisheries' governance must be improved to protect diminishing fish stocks and ensure the livelihood of Africa's fishermen for generations to come.

Interwoven cultural aspects

The oceans and Africans' interaction with them made it possible to create a maritime cultural landscape, evident in diverse aspects ranging from physical structures to language and attitudes. However, cultural values and beliefs form part of the historical socio-economic development processes,[91] which is important for understanding societies in sub-Saharan Africa.

Although the Swahili culture is associated with western Indian Ocean maritime trade, it has an African heritage which stimulated unique economic, political and urban development for centuries.[92] Swahili settlements have a unique building heritage, as architecture evolved from African mud and wattle structures with palm leaf fronds for roofing to complex stone-built towns. A common element of Swahili port towns is their urbanized nature. Essentially, two types of urban settlement emerged: a linear structure along the coastline, or a nucleus structure with a mosque in the centre and the settlement around it.[93]

Fascinating examples of the architectural and spatial splendour of old Swahili coastal settlements exist. Early seafarers report neatly plastered flat-roofed stone houses, up to three storeys high, in densely populated towns with narrow alleys and wide streets with irrigation channels for fruit and vegetable gardens.[94] The historic town of Lamu, one of the oldest and best-preserved examples, covers about 16 hectares along the coastline. Most of its buildings are constructed from coral stone and mangrove timber, and although evidence of earlier settlements exists Lamu has been continually inhabited for more than 700 years.[95] Even though parts of Kilwa are still unexcavated its extensive ruins are proof of an imposing and thriving port city occupied from the ninth to the nineteenth century. Among its many commercial and religious buildings constructed from coral stone and lime mortar, is the Great Mosque (built between the eleventh and thirteenth centuries) with its stone columns and decorated cupolas and vaults.[96]

Maritime contact also made it possible for settlements to share building technologies. Lamu in the north and Kilwa in the south, for example, share similarities in the layout of the buildings and decor. Another shared Swahili architectural feature is the beautifully carved doors. Yet despite such similarities, each town might have unique decorative aspects, such as the floral elements adopted in Lamu.[97]

The African coastline is dotted with palaces, castles and fortifications in various styles, created in different periods and diverse in origin. These include many iconic structures such as the ruins of the fourteenth-century palace fortress of Husuni Kubwa at Kilwa[98] and the Sultan's palace in Stone Town, Zanzibar. Many of the castles along the coast of Africa regularly changed hands. Well-known Elmina Castle, that was important to the slave trade and as a military outpost, was captured by the Dutch in 1637, only to be sold to the British later. Although these forts no longer face such threats, new threats in the form of rising temperatures and sea levels could cause the erosion of these controversial but culturally important buildings.[99]

Maritime interaction and the importance of the Cape of Good Hope as a 'tavern of the seas' changed the history of South Africa. The Dutch East India Company (Verenigde Oost-Indische Compagnie – VOC) established a settlement at the Cape in 1652. It was not a trading post, but its location between the Indian and Atlantic Oceans astride the long sea route that carried enormous riches from the East, made it an ideal refreshment outpost for ships and a position of military strategic importance. As the volume of shipping required substantial supplies of meat, grain, fresh produce and wine, the VOC allocated farms to free burgers, resulting in greater portions of the interior falling under the sway of the ever-expanding agrarian settlement and driving the Khoi herders from their ancestral grazing.

The Dutch created port infrastructure and warehouses in Cape Town and Simon's Town to supply shipping and enhance the security of the Cape sea route. As the settlement and its immediate hinterland had to be defended against powerful European competitors, the Dutch created a substantial military infrastructure. Construction of the key fortification in Cape Town (a large pentagonal stone castle that survived the centuries intact) commenced in 1666 and a series of defensive fortifications were added around the Cape Peninsular.[100] After the British conquered the Cape, they enhanced this infrastructure and also created a naval base in Simon's Town.

The fact that the Cape was part of the greater VOC empire stimulated the development of a unique and complex new society, comprising the indigenous population, European settlers (mainly Dutch, German and French), as well as banished nobles, slaves and artisans from Madagascar, Africa, Malaysia, Bengal, Ceylon, India and Indonesia. Christianity and Islam were introduced, and by the end of the VOC era a new language (now called Afrikaans) had evolved at the Cape.

Maritime contact and colonialism influenced language development and resulted in the prevalence of European languages in the region. Although the exact origins of the Swahili language is obscure,[101] it could be regarded as a fusion of mostly Bantu African language traditions, influenced by Arab, Indian and European languages (specifically Portuguese), among others.[102] The Swahili have borrowed many terms relating to ships and shipbuilding, geographical features and the exploitation of marine resources from Arabic and later from Portuguese, and even from Malay or Javanese.[103] As language is also a vehicle for ideas, the Swahili made it possible to have common communication across the region on matters maritime or matters of business. Swahili is still widely spoken as a lingua franca along the east coast of Africa, the Great Lakes up to the Democratic Republic of the Congo, and from the Comoro Islands to Madagascar.[104]

There are many references to the sea, sea creatures and boats in African folklore, songs, literature and oral poetry. In West Africa these include the story of Anansi, or the story about why the sun and the moon live in the sky.[105] The long history of seafaring around the Horn of Africa resulted in considerable oral poetry on maritime matters, and fabled stories about love, stormy seas and trade, such as the tragic story of Zeïla.[106] There are many stories and fables relating to the sea in Afrikaans and a common folk song 'Daar kom die Alibama' ('the Alibama is coming') is linked to the visits of the Confederate raider CSS *Alabama* to the Cape in 1863 and 1864.

Great historic shocks affect cultural evolution. The trans-Atlantic and Indian Ocean slave trade fed subsequent distrust, as a 'negative relationship between an individual's trust in others and the intensity of the slave trade among their ethnic group in the past'[107] could be identified. The slave trade had an impact on cultural norms and caused a 'deterioration of domestic institutions', making people 'less trusting of others today'.[108] The historic legacy of Arab-led slavery is complicated and is often 'hushed-up because of the embarrassing reaction it generates', yet it affects the 'quests for Arab and African unity'.[109]

Conclusion

The developing and persistent nature of culture indicates that history is central to its evolution and remains relevant.[110] The cultural-historic and maritime experience of sub-Saharan Africa is vast due to its geographic and nautical diversity, dissimilar interaction patterns between people and the ocean, external influences, and the diverse socio-economic character of coastal communities. While there were specific differences between the maritime cultures of western and eastern Africa, as noted above, the entire sub-Saharan African coastline and its hinterlands were irrevocably affected and changed by the trade, cultural influence, power, wealth and disruption inherent to the interaction with others from across the oceans.

The evolving maritime story of sub-Saharan Africa has shown that the length of history is not necessarily of the greatest importance, but rather the intensity of the impact it made on people and societies. This holds true for the tragic history of the maritime slave trade that was conducted along the east coast of Africa as well as the trans-Atlantic slave trade. The catastrophic 'shock' associated with it has reverberated over centuries; its memory is alive and embedded in the African consciousness.

The question beckons: does sub-Saharan Africa have a maritime culture? Yes, it does. But, due to the region's unique maritime histories it is multi-faceted with distinctive characteristics associated with different regions. Yet the sea linked Africans together, albeit at times by agency. As the story of sub-Saharan Africa and the sea is unending, in order to sustain, develop and prosper from the oceans, the region needs to value its maritime past and culture, and continue to forge a new maritime future and culture that is uniquely African.

The maritime interests of the region are considerable; its oceans and ports are not important merely for trade and as a source of food and energy, but historically they have connected the continent to the rest of the world. Recent governments and institutions in sub-Saharan African countries have paid insufficient attention to the oceans, yet today the vast potential of the blue economy is central to economic growth and development and will remain a key element in future development. Maritime interaction and 'across the sea' should not only imply other regions of the world, but the legacy of considerable coastal and intercontinental trade must be reclaimed. And, associated with it, strong regional cooperation in governance, environmental care and security must contribute to collectively charting a sustainable development course.

This close association with the oceans has expressively influenced the historic and cultural evolution of sub-Saharan Africa. Laws, statues and business are not enough to sustain and develop its maritime interests. African leaders and citizenry must 'think maritime' and regard their maritime heritage as more than just a matter of business. Cherishing the region's maritime past and stimulating the maritime culture that has evolved over centuries is an inherent part of creating a continent that will benefit not only from the blue economy, but that will also ensure the survival of the ocean environment for generations to come.

Notes

1 'A man who has been to sea has had a bewildering experience which will be difficult for him to explain' – oral poetry from the Horn of Africa region. Galaal, 'Historical Relations', 25.
2 *The Cambridge English Dictionary.*
3 Nunn, 'Culture', 112–17.
4 Aderibigbe, 'Religious Traditions', 7.
5 Davidson, *Africa*, 44–48; and Aderibigbe, 'Religious Traditions', 8–10.
6 SAFCEI, 'Faiths: African'.
7 Bernard, 'Ecological Implications', 148–49.
8 Wigington, 'Yoruba Religion'.
9 Badejo, 'Olokun', 489.
10 Wigington, 'Yoruba Religion'.
11 Adjei and Sika-Bright, 'Traditional', 2.
12 Ibid., 16; and Akyeampong, 'Indigenous Knowledge', 177.
13 Mollat, 'Historical Contacts', 54.
14 Van Aswegen, *Afrika*, 108.
15 Seland, 'Network Approach', 192–93.
16 *Periplus*; and Seland, 'Network Approach', 195 and 204.
17 Van Aswegen, *Afrika*, 105–06.
18 Harris, *Sea Ports*, 1–11.
19 Chittick, 'East Africa', 15; Mollat, 'Historical Contacts', 52–53; and Allen, 'Proposal', 143 and 161.
20 BBC, 'Coast and Conquest'.
21 Mollat, 'Historical Contacts', 55.
22 De Visse, 'Inter-continental', 658–59.
23 Pollard and Ichumbaki, 'Ports and Harbors', 459, 470 and 485.
24 Henriksen, *Mozambique*, 23–24.
25 Chittick, 'East Africa', 18.
26 Pollard and Ichumbaki, 'Ports and Harbors', 462, 467 and 485.
27 Salim, 'East African', 215–21.

28 Kake, 'Slave Trade', 167–68; and Dima, 'African Diaspora', 19–20.
29 Vellut, 'The Congo Basin', 306.
30 Kake, 'Slave Trade', 167.
31 Daget, 'The Abolition', 88.
32 Kake, 'Slave Trade', 167.
33 Daget, 'The Abolition', 88.
34 Phiri, 'Reaching', 64; and Bankie and Mchombu, *Pan-Africanism*, 20 and 229.
35 Davidson, *Africa*, 129.
36 Mollat, 'Historical Contacts', 53–54.
37 Kriger, *Making Money*, 9.
38 De Visse, 'Inter-continental', 667; and Van Aswegen, *Afrika*, 121–23.
39 Niane, *Africa*, 325–26, 667.
40 Kriger, *Making Money*, 10.
41 Davidson, *Africa*, 57; and Niane, *Africa*, 7–8.
42 Kriger, *Making Money*, 34–35.
43 Ibid., 186.
44 Modupe *et al.*, 'Textile Industry', 282–83 and 287.
45 African Studies, 'Transatlantic Slave Trade'; Beckles, *Slave Voyages*, 94–95; and Domingues da Silva and Misevich, 'Atlantic Slavery', 1–2.
46 Dima, 'African Diaspora', 22.
47 UNESCO, *Slave Route.*
48 Davidson, *Africa*, 62.
49 Beckles, *Slave Voyages*, 4–9.
50 Davidson, *Africa*, 63–4.
51 Anti-Slavery International, *Cocoa Industry*, 5.
52 Davidson, *Africa*, 64–65.
53 Ibid.
54 Ocheni and Nwankwo, 'Analysis of Colonialism', 53.
55 Hbek, 'Emergence of the Fätimids', 21–24; and Ogot, *Africa*, 445.
56 Ogot, *Africa*, 386, 757 and 863.
57 *Africa Times*, 'Global Piracy'.
58 Potgieter and Schofield, 'Poverty, Poaching and Pirates', 105–06.
59 Breen and Lane, 'Archaeological Approaches', 477–78.
60 Van Aswegen, *Afrika*, 134–35 and 208–11.
61 Reynolds, *Command of the Sea*, 419.
62 Tymowski, 'Death and Attitudes', 790–94.
63 Crowder, *West African Resistance*, 27–28; and Reynolds, *Command of the Sea*, 343.
64 Strachan, *European Armies*, 77–88.
65 UNESCO, *Historical Relations*, 166.
66 De Visse, 'Inter-continental', 655.
67 Allen, 'Proposal', 149; Chittick, 'East Africa', 15; and Breen and Lane, 'Archaeological Approaches', 472.
68 Chittick, 'East Africa', 14.
69 Pollard, 'Safeguarding Swahili Trade', 474.
70 Pollard and Ichumbaki, 'Ports and Harbors', 462.
71 Pollard, 'Safeguarding Swahili Trade', 458–59, 472–75.
72 Pollard and Ichumbaki, 'Ports and Harbors', 467.
73 Pollard and Kinyera, 'The Swahili Coast', 933.
74 Matveiev, 'Swahili Civilization', 457; and De Visse, 'Inter-continental', 657.
75 Sutton, 'East Africa', 555–56; Breen and Lane, 'Archaeological Approaches', 477.
76 Pollard and Ichumbaki, 'Ports and Harbors', 467.
77 Ibid., 467–68.

78 Mollat, 'Historical Contacts', 54.
79 Pollard and Ichumbaki, 'Ports and Harbors', 464.
80 Person, 'The Coastal Peoples', 306 and 310.
81 Breen and Lane, 'Archaeological Approaches', 469–70.
82 FAO, *State of World Fisheries.*
83 African Union, *Maritime Strategy*, 8.
84 Breen and Lane, 'Archaeological Approaches', 472.
85 Matveiev, 'Swahili Civilization', 456.
86 Breen and Lane, 'Archaeological Approaches', 475–76.
87 Akyeampong, 'Indigenous Knowledge', 175.
88 Harms, 'Fishing', 147 and 150.
89 Akyeampong, 'Indigenous Knowledge', 174–76.
90 Ibid., 180.
91 Nunn, 'Culture', 122.
92 Pollard and Ichumbaki, 'Ports and Harbors', 459 and 467.
93 BBC, 'Coast and Conquest'.
94 Van Aswegen, *Afrika*, 108.
95 UNESCO, 'Lamu Old Town'.
96 Matveiev, 'Swahili Civilization', 469–71.
97 BBC, 'Coast and Conquest'.
98 Matveiev, 'Swahili Civilization', 470–71.
99 Fagotto, 'West Africa'.
100 Potgieter, 'Die strategiese waarde', 176–79.
101 UNESCO, *Historical Relations*, 178.
102 BBC, 'Coast and Conquest'.
103 Blench, 'Two Vanished', 282.
104 UNESCO, *Historical Relations*, 178; and Blench, 'Two Vanished', 274–75.
105 Baker and Sinclair, *West African.*
106 Galaal, 'Historical Relations', 23–30.
107 Nunn, 'Culture', 115.
108 Ibid.
109 Bankie, 'Arab Slavery', 78.
110 Nunn, 'Culture', 117.

Bibliography

Aderibigbe, Ibigbolade S. 'Religious Traditions in Africa: An Overview of Origins, Basic Beliefs, and Practices'. In *Contemporary Perspectives on Religions in Africa and the African Diaspora*, edited by Ibigbolade Aderibigbe and Carolyn Medine, 7–29. New York: Palgrave Macmillan, 2015.

Adjei, Joseph K. and Solomon Sika-Bright. 'Traditional Beliefs and Sea Fishing in Selected Coastal Communities in the Western Region of Ghana'. *Ghana Journal of Geography* 11, no. 1 (2018): 1–19. https://dx.doi.org/10.4314/gjg.v11i1.1.

Africa Times. 'Global Piracy Down, But Gulf of Guinea Sees a Spike'. *Africa Times.* 15 January 2020. https://africatimes.com/2020/01/15/global-piracy-down-but-gulf-of-guinea-sees-a-spike/.

African Studies Centre Leiden. *Dutch Involvement in the Transatlantic Slave Trade and Abolition.* Web dossiers. Accessed 7 April 2020. www.ascleiden.nl/content/webdossiers/dutch-involvement-transatlantic-slave-trade-and-abolition.

African Union. *2050 Africa's Integrated Maritime Strategy.* Addis Ababa: Africa Union, 2013.

Akyeampong, Emmanuel. 'Indigenous Knowledge and Maritime Fishing in West Africa: The Case of Ghana'. In *Indigenous Knowledge Systems and Sustainable Development: Relevance for Africa*, edited by Emmanuel Boon and Luc Hens, 173–182. Delhi: Kamla-Raj, 2007.

Allen, J.de V. 'A Proposal for Indian Ocean Studies'. In *Historical Relations Across the Indian Ocean, UNESCO Meeting of Experts, Port Louis, July 15–19, 1974*, 137–152. Paris: UNESCO, 1980.

Anti-Slavery International. *The Cocoa Industry in West Africa: A History of Exploitation*. London: Anti-Slavery International, 2004. www.antislavery.org/wp-content/uploads/2017/01/1_cocoa_report_2004.pdf.

Baker, William and Cecilia Sinclair. *West African Folk-Stories*. Chapel Hill, NC: Yesterday's Classics, 2007.

Badejo, Diedre L. 'Olokun'. In *Encyclopedia of African Religion*, edited by Molefi Kete Asante and Ama Mazama, 489. Sage Online, 2009. http://dx.doi.org/10.4135/9781412964623.

Bankie, B. F. 'Arab Slavery of Africans in the Afro-Arab Borderlands'. *African Renaissance* 1, no. 1 (2004): 78–81.

Bankie, B. F. and K. Mchombu, eds. *Pan-Africanism/African Nationalism: Strengthening the Unity of Africa and its Diaspora*. Asmara: The Red Sea Press, 2008.

Beckles, Hilary M. *Slave Voyages: The Transatlantic Trade in Enslaved Africans*. Paris: UNESCO, 2002.

Bernard, Penny S. 'Ecological Implications of Water Spirit Beliefs in Southern Africa: The Need to Protect Knowledge, Nature, and Resource Rights'. *USDA Forest Service Proceedings, RMRS-P-27* (2003): 148–154. www.fs.fed.us/rm/pubs/rmrs_p027/rmrs_p027_148_154.pdf?origin=publication_detail.

BBC. 'Coast and Conquest'. *History of Africa*, Series 2, Episode 3. Broadcasted 7 December 2019. www.bbc.co.uk/programmes/m000cpcv.

Blench, Roger. 'Two Vanished African Maritime Traditions and a Parallel from South America'. *Afr Archaeol Rev* 29 (2012): 273–292. https://doi.org/10.1007/s10437-012-9115-y.

Breen, Colin and Paul J.Lane. 'Archaeological Approaches to East Africa's Changing Seascapes'. *World Archaeology* 35 no. 3 (2003): 469–489.

The Cambridge English Dictionary. Accessed 10 January 2020. https://dictionary.cambridge.org/dictionary/english/.

Chittick, Neville. 'East Africa and the Orient: Ports and Trade before the Arrival of the Portuguese'. In *Historical Relations across the Indian Ocean, UNESCO Meeting of Experts, Port Louis, July 15–19, 1974*, 13–22. Paris: UNESCO, 1980.

Crowder, Michael, ed. *West African Resistance: The Military Response to Colonial Occupation*. London: Hutchinson, 1978.

Daget, S. 'The Abolition of the Slave Trade'. In *Africa in the Nineteenth Century until the 1880s, General History of Africa* VI, edited by J. F. Ade Ajayi, 64–89. Paris: UNESCO, 1989.

Davidson, Basil. *Africa in Modern History: The Search for a New Society*. London: Penguin Books, 1978.

De Visse, J. 'Africa in Inter-continental Relations'. In *Africa from the Twelfth to the Sixteenth Century, General History of Africa* IV, edited by D. T. Niane, 635–672. Paris: UNESCO, 1984.

Dima, S. J. 'The Challenges Faced by the Global African Diaspora'. In *Pan-Africanism/African Nationalism*, edited by B. J. Bankie, 19–29. Asmara: The Red Sea Press, 2008.

Domingues da Silva, Daniel and Philip Misevich. 'Atlantic Slavery and the Slave Trade: History and Historiography'. *African History. Oxford Research Encyclopedias*, 2018. https://doi.10.1093/acrefore/9780190277734.013.371.

Fagotto, Matteo. 'West Africa Is Being Swallowed by the Sea'. *Foreign Policy*, 21 October 2016. https://foreignpolicy.com/2016/10/21/west-africa-is-being-swallowed-by-the-sea-climate-change-ghana-benin/.

Food and Agriculture Organization of the United Nations (FAO). *The State of World Fisheries and Aquaculture 2020: Sustainability in Action*. Rome: FAO, 2020. www.fao.org/state-of-fisheries-aquaculture.

Galaal, Musha. 'Historical Relations between the Horn of Africa and the Persian Gulf and the Indian Ocean Islands through Islam'. In *Historical Relations across the Indian Ocean, UNESCO Meeting of Experts, Port Louis, July 15–19, 1974*, 30–33. Paris: UNESCO, 1980.

Harms, Robert. 'Fishing and Systems of Production: The Precolonial Nunu of the Middle Zaïre'. *Cahiers des Sciences Humaines* 25, no. 1–2 (1989): 147–158. https://horizon.documentation.ird.fr/exl-doc/pleins_textes/pleins_textes_4/sci_hum/30752.pdf.

Harris, Lynn, ed. *Sea Ports and Sea Power: African Maritime Cultural Landscapes*. Cham: Springer, 2017.

Hbek, J. 'The Emergence of the Fätimids'. In *Africa from the Seventh to the Eleventh Century, General History of Africa* III, edited by M. Elfasi, 1–30. Paris: UNESCO, 1988.

Henriksen, Thomas. *Mozambique: A History*. London: Rex Collings, 1978.

Kake, I. B. 'The Slave Trade and the Population Drain from Black Africa to North Africa and the Middle East'. In *The African Slave Trade from the Fifteenth to the Nineteenth Century, UNESCO Meeting of Experts, Port-au-Prince, January 31 to February 4, 1978*. Paris: UNESCO, 1979.

Kriger, C. E. *Making Money: Life, Death, and Early Modern Trade on Africa's Guinea Coast*. Athens: Ohio University Press, 2017.

Matveiev, V. V. 'The Development of Swahili Civilization'. In *Africa from the Twelfth to the Sixteenth Century, General History of Africa* IV, edited by D. T. Niane, 455–481. Paris: UNESCO, 1984.

Mkenda, Bakanja. 'Environmental Conservation Anchored in African Cultural Heritage'. *New People Magazine* 125, 1 April 2010. www.africafiles.org/article.asp?ID=23343.

Modupe, Adu F., Ajayi Adeyinka and Johnson O. Aremu. 'Textile Industry in Yorubaland: Indigenous Knowledge and Modernity in the Era of Globalisation'. *Advances in Social Sciences Research Journal* 5, no. 4 (2018): 282–292.

Niane, D. T., ed. *Africa from the Twelfth to the Sixteenth Century, General History of Africa* IV. Paris: UNESCO, 1984.

Mollat, Michel. 'Historical Contacts of Africa and Madagascar with South and South-East Asia: The Role of the Indian Ocean'. In *Historical Relations across the Indian Ocean, UNESCO Meeting of Experts, Port Louis, July 15–19, 1974*, 45–60. Paris: UNESCO, 1980.

Nunn, Nathan. 'Culture and the Historical Process'. *Economic History of Developing Regions* 27, S1 (2012): 108–126. https://doi.org/10.1080/20780389.2012.664864.

Ocheni, Stephen and Basil Nwankwo. 'Analysis of Colonialism and Its Impact in Africa'. *Cross-Cultural Communication* 8, no. 2 (2012): 46–54. https://doi.10.3968/j.ccc.1923670020120803.1189.

Ogot, B. A., ed. *Africa from the Sixteenth to the Eighteenth Century, General History of Africa* V. Paris: UNESCO, 1992.

Periplus Maris Erythraei. Translated as *Voyage around the Erythraean Sea*. Middle of the first century CE. https://depts.washington.edu/silkroad/texts/periplus/periplus.html.

Person, Y. 'The Coastal Peoples: From Casamance to the Ivory Coast Lagoons'. In *Africa from the Twelfth to the Sixteenth Century, General History of Africa* IV, edited by D. T. Niane, 302–323. Paris: UNESCO, 1984.

Phiri, Isabel A. 'Reaching the Champions of Social Justice'. *The Ecumenical Review* 72, no.1 (2020): 62–72.

Pollard, Edward. 'Safeguarding Swahili Trade in the Fourteenth and Fifteenth Centuries: A Unique Navigational Complex in South-East Tanzania'. *World Archaeology* 43, no. 3 (2011): 458–477. http://dx.doi.org/10.1080/00438243.2011.608287.

Pollard, Edward and Okeny C.Kinyera. 'The Swahili Coast and the Indian Ocean Trade Patterns in the 7th–10th Centuries CE'. *Journal of Southern African Studies* 43, no. 5 (2017): 927–947. https://doi.org/10.1080/03057070.2017.1345266.

Pollard, Edward and Elgidius B.Ichumbaki. 'Why Land Here? Ports and Harbors in Southeast Tanzania in the Early Second Millennium AD'. *The Journal of Island & Coastal Archaeology* 12 (2017): 459–489.

Potgieter, Thean. 'Die strategiese waarde van die Kaap en die VOC se verdedigingstelsel'. In *Die VOC aan die Kaap 1652–1795*, edited by Con De Wet, Leon Hatting and Jan Visagie, 160–180. Pretoria: Protea, 2016.

Potgieter, Thean and Clive Schofield. 'Poverty, Poaching and Pirates: Geopolitical Instability and Maritime Insecurity off the Horn of Africa'. *Journal of the Indian Ocean Region* 6, no. 1 (2010): 86–112. https://doi.org/10.1080/19480881.2010.489673.

Reynolds, Clark. *Command of the Sea: The History and Strategy of Maritime Empires*. London: Morrow, 1976.

Southern African Faith Communities' Environment Institute (SAFCEI). *Faiths: African Traditional*. Cape Town: SAFCEI, 2018. https://safcei.org/project/african-traditional/.

Salim, A. I. 'The East African Coast and Hinterland, 1800–45'. In *Africa in the Nineteenth Century until the 1880s, General History of Africa* VI, edited by J. F. Ade Ajayi, 211–233. Paris: UNESCO, 1989.

Seland, Eivind. 'The *Periplus* of the Erythraean Sea: A Network Approach'. *The Asian Review of World Histories* 4 (2016): 191–205. https://doi.10.12773/arwh.2016.4.2.191.

Strachan, Hew. *European Armies and the Conduct of War*. Reprint. London: Routledge, 1993.

Sutton, J. E. G. 'East Africa before the Seventh Century'. In *Ancient Civilizations of Africa, General History of Africa* II, edited by G. Mokthar, 568–593. Paris: UNESCO, 1981.

Tymowski, Michal. 'Death and Attitudes to Death at the Time of Early European Expeditions to Africa (15th Century)'. *Cahiers d'études africaines* 215 (2014): 787–811. https://journals.openedition.org/etudesafricaines/17843.

United Nations Educational, Scientific and Cultural Organization (UNESCO). *Historical Relations across the Indian Ocean. UNESCO Meeting of Experts, Port Louis, July 15–19, 1974*. Paris: UNESCO, 1980.

United Nations Educational, Scientific and Cultural Organization (UNESCO). *The Slave Route*. Paris: UNESCO, n.d. Accessed 12 May 2020. www.unesco.org/new/fileadmin/MULTIMEDIA/HQ/CLT/pdf/MapSlaveRoute.pdf.

United Nations Educational, Scientific and Cultural Organization (UNESCO). 'Lamu Old Town'. *World Heritage List*, Paris: UNESCO, 2001. https://whc.unesco.org/en/list/1055/.

Van Aswegen, H. J. *Geskiedenis van Afrika*. Pretoria: Academica, 1980.

Vellut, J. L. 'The Congo Basin and Angola'. In *Africa in the Nineteenth Century until the 1880s, General History of Africa* VI, edited by J. F. Ade Ajayi, 294–324. Paris: UNESCO, 1989.

Wigington, Patti. 'Yoruba Religion: History and Beliefs'. *Learn Religions*, 11 February 2020. www.learnreligions.com/yoruba-religion-4777660.

2 Tourism, ecosystems, biodiversity and threats

Angela M. Lamptey, PhD

Introduction

Sub-Saharan Africa supports extremely diverse ecological communities across its eight terrestrial biomes (which include forests, savannahs, woodlands, grasslands, scrublands, deserts and mangroves) as well as multiple freshwater and marine ecosystems. The region's complex climate, geology and history have contributed to developing its exceptional biodiversity. Tourism is playing an increasingly large role in the region's economies (although the coronavirus (COVID-19) pandemic has slowed such growth for the short term).

Since probably 3500 BCE, the Saharan and sub-Saharan regions of Africa have been separated by the extremely harsh climate of the sparsely populated Sahara, forming an effective barrier interrupted by only the Nile in Sudan, albeit that the Nile was blocked by the river's cataracts. The Sahara pump theory explains how flora and fauna (including *Homo sapiens*) left Africa to penetrate the Middle East and beyond. African pluvial periods are associated with a Saharan wet phase, during which larger lakes and more rivers existed.[1]

Terrestrial biomes/ecosystems in sub-Saharan Africa

Much of the African continent encompasses the Afro-tropical eco-region, which is separated from other eco-regions by the Indian Ocean to the east, the Atlantic Ocean to the west and the Sahara Desert to the north. These major geographic features have acted as barriers to movement since the African continent first took its current shape, enabling species and ecosystems characteristic of the region to evolve in relative isolation from those of other eco-regions. The Afro-tropical eco-region can be further subdivided into eight terrestrial biomes, each with its own distinct climate, geology and biota.[2]

The region's topographic complexity, the diversity of biomes and the multiple ecological transition zones between the different biomes have given rise to a rich biodiversity.[3]

Tropical and subtropical savannahs and grasslands including woodlands, bushlands, thickets and semi-arid drylands form the largest biome in sub-Saharan Africa. They are maintained by fire and grazing. East and Southern Africa's miombo and mopane savannah-woodland are included in this ecosystem. The deserts and arid scrublands have evaporation levels exceeding precipitation, generally with rainfall less than 250 mm per year. Tropical moist forests are lowland broadleaf ecosystems with near-continuous canopies that run as a broad band across equatorial Africa. This biome is characterized by high rainfall (more than 2 m per year), low variability in temperatures and very high species diversity. Montane grasslands and scrublands are patchily distributed biome that occur at altitudes greater than 800 m and have enough rainfall for a variety of grasses to thrive. A Mediterranean scrubland ecosystem of limited extent, better known as the Fynbos or Cape Floristic Region, is situated at Africa's south-western tip. Characterized by hot dry summers and cool moist winters, it contains one of Earth's richest concentrations of endemic plant species. Flooded grasslands and savannahs include marshes and shallow lakes that are periodically flooded by water that can be fresh, brackish or hypersaline. When flooded, these areas host some of the largest waterbird congregations in the region. Tropical dry forests are highly restricted forest types that can be found in western Zambia and adjacent Angola, as well as in Cabo Verde. While these areas may receive high rainfall, they are characterized by seasonal droughts that can last several months. Mangrove ecosystems are coastal wetlands of tropical climates characterized by distinctive woody plants with aerial roots that can tolerate saltwater. Typically associated with intertidal zones and muddy bottoms, mangroves provide nursery grounds for many aquatic animal species.[4] [5]

The Cape Floristic Region is home to the greatest concentration of non-tropical endemic species in the world, including well-known plant genera like *Protea* and *Erica*. The Succulent Karoo, directly north of the Cape Floristic Region, may be the most floristically rich desert in the world.[6] Africa has deservedly received international acclaim for these and many other natural wonders. Prominently, more than 37 sites in sub-Saharan Africa have already been recognized as natural World Heritage sites. One such site is also Africa's oldest national park, Virunga National Park in eastern Democratic Republic of the Congo (DRC), which contains at least 218 mammal and 706 bird species.[7]

The variety of biomes present in sub-Saharan Africa is the result of variable geology and a long history of changes in climate and ecological communities. For example, when the Earth's climate was warmer, tropical moist forests were more widely distributed. As the planet cooled during glacial periods, forests contracted and became fragmented while grasslands expanded; some new biomes developed as the climate changed and species moved around. Even today, biome boundaries are still shifting: for example, over the past few decades the boundary between the Sahara Desert and the Sahel has shifted by hundreds of kilometres southward.[8] The development, fragmentation and movement of these and other biomes, as well as the influence of major dispersal barriers, such as large rivers and mountain ranges, have stimulated speciation, as different populations became specialized to conditions that were restricted to their particular elevations or on certain sides (wet or dry, sunny or shady) of mountain ranges.

Aquatic and coastal biomes/ecosystems

In addition to these terrestrial biomes, sub-Saharan Africa also contains several aquatic biomes. Prominent freshwater biomes include several large rivers along with their headwaters and deltas, numerous small rivers, multiple large and small lakes, as well as a variety of wetland ecosystems such as swamps, bogs and salt marshes.[9] Prominent marine biomes include tropical coral reefs along Africa's east coast, as well as temperate continental shelves and seas along the coastlines of South Africa and Namibia. There are also several important oceanic upwelling areas; these include the tropical Gulf of Guinea upwelling along West Africa, and the Benguela upwelling ecosystem along Africa's south-west coast.[10]

Biodiversity

Due to this dynamic geological, climatic and environmental history, as well as all the factors that have promoted speciation, sub-Saharan Africa boasts tremendous species richness. The region is very well known for its mammals, particularly its charismatic terrestrial megafauna and other large mammals that attract millions of tourists from all around the world each year. Among the most famous are the 'Big Five' animals such as lions (*Panthera leo*, VU), savannah elephants (*Loxodonta africana*, VU), African buffalo (*Syncerus caffer*, NT), African leopards (*Panthera pardus*, VU) and black rhinoceros (*Diceros bicornis*, CR). Other notable mammals include the cheetah (*Acinonyx jubatus*, VU), the fastest mammal on Earth; the Maasai giraffe (*Giraffa camelopardalis tippelskirchii*, VU), the world's tallest mammal; the giant eland (*Tragelaphus derbianus*, VU), the world's largest antelope; and Africa's four species of great ape. Many small mammals are also noteworthy. For example, East Africa's naked mole-rat (*Heterocephalus glaber*, LC) is the world's only mammalian thermo-conformer.[11] The naked mole-rat and Southern Africa's Damaraland mole-rat (*Fukomys damarensis*, LC) are the only known eusocial mammals; like some ants and bees, only one female (the queen) reproduces with one to three breeding males, while all the other colony members are sterile workers.[12]

While Africa's large mammals are a major tourist attraction, the region hosts many other rich and noteworthy wildlife assemblages. With more than 2,100 bird species, 1,400 of them found nowhere else on Earth,[13] the Afrotropics may be the most taxonomically diverse bird region on the planet.[14]

Among the many bird species that call Africa home is the world's largest extant species of bird, the red-necked ostrich (*Struthio camelus camelus*); standing up to 2.74 m tall, it is in dire need of conservation attention.[15] Africa is also home to the world's heaviest extant flying animal, the kori bustard (*Ardeotis kori*, NT), which can weigh over 20 kg.[16] Over 100,000 insects have been described in sub-Saharan Africa,[17] which include the world's smallest butterfly, the dwarf blue (*Oraidium barberae*, LC) of Southern Africa and the aptly named goliath beetle (*Goliathus* spp.), which can be found throughout much of tropical Africa. The region also hosts a great number of noteworthy endemic amphibians and reptiles, including the world's largest frog, the goliath frog (*Conraua goliath*, EN) of Cameroon and Equatorial Guinea and the black mamba (*Dendroaspis polylepis*, LC), arguably the world's most feared snake, which is widespread across Africa's savannahs. Finally, Africa is home to Jonathan the Aldabra giant tortoise (*Aldabrachelys gigantea*, VU); having hatched in 1832, he is considered the oldest living terrestrial animal in the world. A few small and isolated African ecosystems are particularly rich in species. Particularly noteworthy are the Rift Valley lakes, such as Lake Victoria, Lake Malawi and Lake Tanganyika, which hold the richest freshwater fish diversity in the world. For example, nearly 14% of the world's freshwater fish species occur in Lake Malawi (also known as Lake Nyasa). Moreover, over 90% of Lake Malawi's fish species are endemic, and thus found nowhere else on Earth.

The region's plant richness, estimated at over 45,000 species,[18] is also important from a global perspective. Many plant species have high economic value, particularly those that have been domesticated in the region and are now important crops across the world. Primary among these is coffee, second only to tea in worldwide popularity as a beverage, which is native to West and Central Africa (*Coffea robusta*) and Ethiopia (*Coffea arabica*). Other important crops that originated in the Afrotropics include okra, black-eyed peas, watermelon and African oil palm. Conserving the wild genetic diversity of these domesticated plants in their native ranges is important because they may serve as excellent replacements for today's crops that may be less productive in future due to anthropogenic climate change.[19] Others, such as the wide variety of plants utilized in traditional medicine to treat malaria, may one day lead to the development of new antimalarial drugs.[20] Similarly, many plant species also have high evolutionary value. These include relict species that survived previous mass extinction events, such as cycads (*Encephalartos* spp.) and Lazarus species that were once believed to be extinct, such as Seychelles' unique jellyfish tree (*Medusagyne oppositifolia*, CR).

Climate zones and biomes

Sub-Saharan Africa has a wide variety of climate zones or biomes. South Africa and the DRC in particular are considered megadiverse countries, having a dry winter season and a wet summer season. The Sahel extends across all of Africa at a latitude of about 10° to 15° N. Countries that include parts of the Sahara Desert in their northern territories and parts of the Sahel in their southern region include Mauritania, Mali, Niger, Chad and Sudan. The Sahel has a hot semi-arid climate. South of the Sahel, a belt of savanna (the West and East Sudanian savannas) stretch from the Atlantic Ocean to the Ethiopian Highlands. The more humid Guinean and Northern Congolian forest-savanna mosaic lies between the savannas and the equatorial forests. The Horn of Africa has a hot desert climate along the coast, but a hot semi-arid climate is more prevalent in the interior, contrasting with savanna and moist broadleaf forests in the Ethiopian Highlands. Tropical Africa encompasses tropical rainforest stretching along the southern coast of West Africa and across most of Central Africa (the DCR) west of the African Great Lakes. In Eastern Africa, woodlands, savannas and grasslands are found in the equatorial zone, including the Serengeti ecosystem in Tanzania and Kenya. Distinctive Afromontane forests, grasslands and shrublands are found in the high mountains and mountain ranges of Eastern Africa, from the Ethiopian Highlands to South Africa. South of the equatorial forests, the western and southern Congolian forest-savanna mosaics are transition zones between the tropical forests and the miombo woodland belt that spans the continent from Angola to Mozambique and Tanzania. The Namib and Kalahari Deserts lie in south-western Africa and are surrounded by semi-deserts including the Karoo in western South Africa. The Bushveld grasslands lie to the east of the deserts.

Biodiversity conservation efforts

Traditional communities have long held the belief that human beings are physically and spiritually connected to nature, and that communal needs outweigh individual desires. This also can also be extended to natural resources, which were considered communal property that must also be shared with the spirits of the ancestors and future generations. Managing natural resources in this way required strict adherence to customary law systems that imposed controls on the collecting of animal and plant products. Some animals and plants were also worshipped, leading to mythical superstitions and taboos that prohibited the killing of culturally and spiritually important animals, as well as totem species that bonded families and villages together. Customary laws also created Africa's first protected areas, such as royal hunting grounds (areas where kings and traditional chiefs had exclusive hunting rights) and areas of spiritual significance, where access to and harvesting of natural resources were restricted.

This culturally driven system of checks and balances was greatly disrupted by the arrival of European settlers in the seventeenth century. Armed with guns, and giving little thought to sustainability, the earliest colonists killed thousands of animals for food, trophies, sport and profit. Following concerns about declining wildlife populations, particularly at the southern tip of South

Africa, sub-Saharan Africa's first formal environmental legislation was introduced in 1657, followed by the region's first formal environmental law in 1684.[21] Significantly, this first law separated protected species, such as the common hippopotamus (*Hippopotamus amphibious*, VU), from pest species (which at the time included lions). Unfortunately, these early laws and regulations were of little consequence as an increasing number of colonists, lured by the promise of unlimited hunting on unexplored lands, arrived in the region. Consequently, by 1700 populations of every animal over 50 kg located within 200 km from Cape Town were extirpated.[22] These developments also led to Africa's first modern man-caused mammal extinctions. First to disappear was the bluebuck (*Hippotragus leucophaeus*, EX) around 1798. Nearly a century later, in 1871, the Cape warthog (*Phacochoerus aethiopicus aethiopicus*, EX) – more closely related to East Africa's desert warthog (*Phacochoerus aethiopicus delamerei* LC) than to the widespread common warthog (*Phacochoerus africanus*, LC) – disappeared, followed by the quagga (*Equus quagga quagga*, EX) in around 1878 (the last captive individual died in 1883). Elsewhere, the bontebok (*Damaliscus pygargus pygargus*, NT), the Cape mountain zebra (*Equus zebra zebra*, VU), the southern white rhinoceros (*Ceratotherium simum simum*, NT), and the black wildebeest (*Connochaetes gnou*, LC) were all reduced to about a dozen individuals at one or two locations.

Ecosystems – forests in particular – near early European settlements similarly suffered as early colonists perceived them as an 'inexhaustible' supply of fuel and timber. This widespread overharvesting prompted the governor of the Cape Colony in 1778 to appoint its first professional nature conservator, Johann Fredrick Meeding, to exercise some control over deforestation. But, like controls on hunting large mammals, these efforts generally only had a local and temporary impact.

Interest in the formal protection of Africa's biodiversity started to intensify during the nineteenth century. Most of the initial steps were taken in South Africa, which had the largest early colonial settlements and, hence, the most species threatened by human activities. First, in 1822 the Game Law Proclamation introduced hunting license fees and closed seasons for selected species, followed by regulations to protect 'open spaces' in 1846 and forests in 1859. A major step towards ecosystem protection was taken in 1876 with the creation of the Cape Colony's Department of Forests and Plantations, while the appointment of a superintendent of woods and forests in 1881 led to initial efforts towards the scientific management of ecosystems. Then, in 1886, the British government passed the Cape Act for the Preservation of Game (in 1891 extended to other British South African territories), followed by the Cape Forest Act of 1888. The Cape Forest Act played an instrumental role in the proclamation of the Cape Colony's first formally protected areas, namely the Tsitsikamma and Knysna Forest Reserves, in 1888; today, these lands are incorporated into South Africa's Garden Route National Park. These were followed by the appointment of Southern Africa's first formal game warden, H. F. van Oordt, in 1893, to manage Pongola Nature Reserve, proclaimed in 1894. (Pongola was degazetted and converted into agricultural land in 1921 but re-established in 1979). Thereafter, protected areas were established at regular intervals across South Africa, starting with Groenkloof Nature Reserve in February 1895, then Hluhluwe Valley and Umfolozi Junction Game Sanctuaries (today known as the Hluhluwe-iMfolozi Park) in April 1895. (St Lucia Game Reserve, today part of iSimangaliso Wetland Park, was also established sometime in 1895.)

West and Central Africa witnessed the first steps towards formal conservation efforts in 1885, with the establishment of forest reserves to protect valuable timber products.[23] The region's first game reserves were gazetted as early as 1889 in the Belgium Congo (now the DRC) to protect elephants. Unfortunately, these efforts were of little consequence as ivory hunters continued to slaughter the region's elephant populations. It was only after colonial governments raised concerns about declining ivory revenues that the region passed its first formal environmental law in 1892, with the ratification of the Congo Basin Convention to regulate the ivory trade in the French, Portuguese and Belgian territories.[24]

In East Africa, colonial authorities passed the first formal environmental legislations in 1888. These initial laws called for establishing game reserves, hunting quotas for common species, strict protection for breeding females and immature animals, and hunting bans for rare species.[25] While the establishment of protected areas was initially slow, a circular from Lord Salisbury (the British prime minister at the time) in which he called for protected areas and hunting restrictions to prevent large mammal extinctions, prompted the passing of the German East African Game Ordinance of 1896. That same year, East Africa witnessed the proclamation of its first modern protected areas, both in Tanzania: one along the Rufiji river (today included in the Selous Game Reserve), and one west of Mount Kilimanjaro.

Initial laws and regulations to protect Africa's environment were greatly expanded in 1900, with the signing of the Convention on the Preservation of Wild Animals, Birds and Fish in Africa, during the International Conference of the African Colonial Powers held in London, UK. While this convention never came into force (because not enough parties ratified it), several signatories continued to follow the convention's agreements by establishing wildlife reserves. Among the first to act were Ghana and Sierra Leone (both governed by Great Britain at the time), which took their first formal steps towards conserving the environment in 1901. Soon afterwards, in 1903 the Society for the Preservation of Wild Fauna of the Empire (today known as Fauna & Flora International, or FFI) became Africa's first conservation non-governmental organization (NGO).

In 1925 Africa's first national park, the Virunga National Park in the Albertine Rift (today divided into the DRC's Virunga and Rwanda's Volcanoes National Parks) was proclaimed. The following year, South Africa's Sabie Game Reserve (which was originally gazetted in 1898) was renamed and expanded as Kruger National Park. Although most early laws focused on protecting rare and 'valuable' mammals, birds, tortoises and timber forests, the welwitschia (*Welwitschia mirabilis*) was the first African plant to enjoy formal protection after colonial powers ratified the 1933 Convention Relative to the Preservation of Fauna and Flora in the Natural State (often referred to as the London Convention).[26]

Fortunately, African conservation biologists regularly employ a can-do attitude, shown in a long history of resourcefulness in the face of resource constraints. For example, conservationists from all over the region have established non-profit NGOs to facilitate a variety of innovative mechanisms to advance biodiversity conservation. One notable example is the African Parks Network; as of mid-2019 the Network, in partnership with its host governments, was managing 15 national parks in nine countries, covering 10.5 million hectares. Through this collaboration, which includes extensive community engagement and law enforcement, wildlife in several once-declining parks is beginning to prosper. For example, lions were reintroduced to Rwanda in 2016 after a 20-year absence, elephant strongholds in Chad and the DRC have been secured, and populations of threatened large mammals on Zambia's Liuwa Plains have increased by 50% to over 100% in just a few years.[27]

Having witnessed all the social and economic benefits achieved through biodiversity conservation efforts, many local communities have been inspired to take the lead in protecting wildlife on their own lands. For example, community efforts have successfully safeguarded Mali's savannah elephants[28] and Rwanda's mountain gorillas (*Gorilla beringei*, EN)[29] during periods of conflict. Locally managed forest reserves now protect more than 36,000 sq km of land in Tanzania,[30] while conservation efforts in community conserved areas in Kenya have renewed hope for the future of the world's rarest antelope, the hirola (*Beatragus hunter*, CR).[31] These examples have set a positive, enterprising tone that has enabled conservation to play in increasingly prominent role in multiple economies through the creation of job opportunities while also improving Africans' overall quality of life.

Tourism

African tourism comprises only 3% of world tourism and with 20% of the world's area being sub-Saharan Africa, the value of its tourist trade should have the potential to reach approximately US $400 billion. A study by the Africa Group shows that tourism potential in Africa is $250 billion. Tourism currently generates approximately $50 billion of sub-Saharan Africa's gross domestic product (GDP), which is much less than its potential value.[32] In 2018 the industry's GDP was approximately $42.1 billion, with 37.4 million tourist arrivals in 2017, about 1.6 % and 3.0% of the global total, respectively. When it comes to tourism, sub-Saharan Africa is a region with massive potential tempered by multiple structural, institutional, economic and socio-economic challenges. According to data from the World Travel and

Tourism Council (WTTC), the region's direct travel and tourism GDP is forecast to grow by 60% from 2018–29 (although this estimate was calculated prior to the outbreak of COVID-19 in early 2020). As such, tourism is expected to play a crucial role in the region's development. However, in order to maximize future gains, regional economies will need to improve their competitiveness. The 2019 Travel and Tourism Competitiveness Report (TTCR) indicates that the region has an enormous opportunity to take advantage of its abundant natural resources and should build its competitiveness through its well-documented successes in increased connectivity. In 2013 the East African Community lined up several projects that were intended to boost tourism earnings from $7 billion to $16 billion by 2020 (this figure was not realized due to the COVID-19 pandemic). It was envisaged that tourist numbers would double from 5 million to 10 million annually. The planned investments had been projected at $3.95 billion by 2020, up from $1.65 billion in 2014. The TTCR highlighted the steps that could be taken to make better use of sub-Saharan Africa's rich natural resources to generate nature-based tourism (popularly known as eco-tourism). About 50% of the 34 regional economies covered by the report are among the upper half of top scorers for natural resources, with Tanzania, South Africa and Kenya among the top 20%.

Sub-Saharan Africa accounts for nearly 17% of UNESCO's World Heritage natural sites and according to the International Union for Conservation of Nature, countries in sub-Saharan Africa have nearly 15% more known species than others. Sub-Saharan Africa's travel and tourism market is very small. In general, with the majority of the region's economies classified as either low or lower-middle income, sub-Saharan Africa lacks the robust middle class and economic resources required to generate intra-regional travel and tourism investment at the same level as other parts of the world, although both areas are demonstrating steady growth. In particular, the current lack of investment means that the region has the least developed infrastructure in the world. Despite this, the region is the least competitive in the world with regard to travel and tourism, with only Mauritius, South Africa and Seychelles ranking in the upper half of the Travel and Tourism Competitiveness Index. This gap is characterized by issues such as an unfavourable business environment, health, safety and hygiene concerns, low ICT readiness and severely underdeveloped infrastructure. In addition, despite the sector's potential and importance to economic development, regional economies have yet to develop strategies to capitalize on this. The range of challenges facing the continent reinforces the notion that competitiveness and development of travel and tourism requires an integrated, holistic approach. In addition, given sub-Saharan Africa's potential for travel and tourism and the broad range of factors that need consideration, the industry can be used as a rallying point around which policymakers and other coastal travel and tourism stakeholders push for improvements in areas ranging from structural reform to infrastructure development.

Maritime security significantly affects the potential of maritime tourism industries in some African states. The peak of piracy in the western Indian Ocean led to a 95% decrease in Kenyan cruise ship tourism (Voice of America). Tourism accounts for about 10% of the GDP of a number of African states such as Seychelles, Cabo Verde and Mauritius.[33]

Threats to biodiversity conservation and tourism

Despite many examples of progress, conservation challenges and conflicts persist across sub-Saharan Africa. As a result, the region lags in several aspects of the safeguarding of its natural heritage. The causes are many and vary by region. Below is a discussion of some of the more prominent impediments to effective conservation action in Africa.

Persistent poverty: There is a direct link between poverty and conservation failure. This is a problem particularly in sub-Saharan Africa, where millions of people live in extreme poverty that is difficult to escape. Faced with hard choices to ensure that there is food on the table, poverty can drive desperate people to illegally collect natural products from protected areas, even though they likely understand the detriment these actions may have on society at large and their own future.[34] [35]

Obstructive mindsets: Colonial Africa has provided many examples showing that conservation activities implemented in an authoritarian manner are bound to fail. Yet authoritarian mindsets continue to impede conservation efforts throughout the region. Work from Guinea-Bissau has shown that authoritarian conservation actions that disempower or displace local communities are more likely to worsen than overcome conservation challenges in post-colonial Africa. Conservation in Africa is as much about people as it is about wildlife.[36]

Weak governance/institutional structures: Africa's natural environment and its people often fall victim to weak governance and institutional structures. It is well known that weak policies, failing governments, corruption, greed and civil conflict hamper conservation efforts and drive biodiversity declines.[37] [38] [39] But even in well-functioning countries, government officials turning a blind eye may enable corporations to cut corners for increased profits at the cost of the environment.[40]

Lack of expertise: Scientific advances depend on increased or updated knowledge. That is also true for conservation biology. Effective conservation depends on local experts who can design and implement monitoring and research projects, apply adaptive management when needed, act as managers and advocates for conservation activities, and increase awareness of the importance of the environment.[41] It is therefore of great concern that conservation in Africa continues to face an enduring skills shortage.[42]

Conflicts of interests: There is always a risk that a wealthy business will threaten a conservation initiative with competing offers that typically include promises of jobs and development.[43] Local people, especially those living in poverty, may find it hard to turn down such attractive counter-offers, even if they recognize that those offers rarely live up to the promises being made to them. Conservation biologists should carefully consider what such offers might look like and factor in how their conservation programmes compete and bring better results for all.

Other threats to biodiversity include deforestation, agriculture, overfishing, hunting and poaching of terrestrial biodiversity, proliferation of invasive/alien species, urbanization, pollution (agriculture, forestry, domestic, industrial and military wastes) and climate change.[44]

Conclusion

Owing to the many challenges that conservation projects continue to face, the list of sub-Saharan African species and ecosystems that are threatened with extinction and destruction continues to grow every year. In a recent assessment, BirdLife International identified 51 Important Bird and Biodiversity Areas in Sub-Saharan Africa, many of them national parks in danger of ecosystem degradation.[45] A United Nations assessment similarly found that the outlook of 12 natural World Heritage sites situated in Sub-Saharan African are 'in danger'.[46] These are substantial and challenging problems that will keep conservation biologists very busy in the future. These problems need to be faced head-on to ensure that future generations will also be able to enjoy the natural treasures and resources the region has to offer.

The blue economy has the potential to make a significant contribution to sub-Saharan Africa's sustainable future growth and development. However, this cannot be realized unless sub-Saharan African governments continue to address maritime security threats and invest in good governance. African states need to strengthen fisheries management and increase investment in domestic fishing industries. The current structure of industry cuts sub-Saharan African economies out of the most valuable portions of the production chain. Security gains in coastal areas, on shore and at sea, could untap huge potential for the development of the tourism industry.

Notes

1 E. M. van Zinderen-Bakker (14 April 1962). 'A Late-Glacial and Post-Glacial Climatic Correlation between East Africa and Europe'. *Nature* 194 (4824): 201–03. Bibcode:1962Natur.194.201V. doi:10.1038/194201a0. S2CID 186244151.
2 N. Burgess, J. D. Hales, E. Underwood and E. Dinerstein 2004. *Terrestrial Ecoregions of Africa and Madagascar: A Conservation Assessment* (Washington, DC: Island Press).
3 D. M. Olson, E. Dinerstein, E. D. Wikramanayake *et al.* 2001. 'Terrestrial Ecoregions of the World: A New Map of Life on Earth'. *BioScience* 51: 933–38. https://doi.org/10.1641/0006-3568(2001)051[0933:TEOTWA]2.0.CO;2.
4 Burgess *et al. Terrestrial Ecoregions of Africa and Madagascar.*
5 Olson *et al.* 'Terrestrial Ecoregions of the World'.

6 R. A. Mittermeier, , P. Robles-Gil, M. Hoffman *et al.* 2004. *Hotspots Revisited: Earth's Biologically Richest and Most Endangered Terrestrial Ecoregions* (Chicago: University of Chicago Press).
7 WHC (World Heritage Committee). 2007. Nomination of Natural, Mixed and Cultural Properties to the World Heritage List – Virunga National Park. Decision 31COM 8B.74 (Christchurch: UNESCO). https://whc.unesco.org/en/decisions/1377.
8 J. A. Foley, M. T. Coe, M. Scheffer *et al.* 2003. 'Regime Shifts in the Sahara and Sahel: Interactions between Ecological and Climatic Systems in Northern Africa'. *Ecosystems* 6: 524–32. https://doi.org/10.1007/s10021-002-0227-0.
9 WWF/TNC. 2013. *Freshwater Ecoregions of the World.* www.feow.org
10 M. D. Spalding, H. E. Fox, G. R. Allen *et al.* 2007. 'Marine Ecoregions of the World: A Bioregionalization of Coastal and Shelf Areas'. *BioScience* 57: 573–83. https://doi.org/10.1641/B570707.
11 R. Buffenstein and S. Yahav. 1991. 'Is the Naked Mole-Rat Heterocephalus glaber an Endothermic yet Poikilothermic?' *Journal of Thermal Biology* 16: 227–32. https://doi.org/10.1016/0306-4565(91)90030-6.
12 J. U. M. Jarvis, M. J. O'Riain, N. C. Bennett *et al.* 1994. 'Mammalian Eusociality: A Family Affair'. *Trends in Ecology and Evolution* 9: 47–51. https://doi.org/10.1016/0169-5347(94)90267-4.
13 I. Sinclair and P. Ryan. 2011. *Birds of Africa South of the Sahara* (Johannesburg: Penguin Random House).
14 C. N. Lotz, J. A. Caddick, M. Forner *et al.* 2013. 'Beyond just Species: Is Africa the Most Taxonomically Diverse Bird Continent?' *South African Journal of Science* 109(5/6): 1–4. https://doi.org/10.1590/sajs.2013/20120002.
15 J. M. Miller, S. Hallager, S. L. Monfort *et al.* 2011. 'Phylogeographic Analysis of Nuclear and mtDNA Supports Subspecies Designations in the Ostrich (Struthio camelus)'. *Conservation Genetics* 12: 423–31. https://doi.org/10.1007/s10592-010-0149-x.
16 J. B. Dunning, Jr. 2008. *CRC Handbook of Avian Body Masses* (Boca Raton: CRC Press).
17 S. E. Miller and L.M. Rogo. 2011. 'Challenges and Opportunities in Understanding and Utilisation of African Insect Diversity'. *Cimbebasia* 17: 197–218
18 R. R. Klopper, L. Gautier, C. Chatelain *et al.* 2007. 'Floristics of the Angiosperm Flora of Sub-Saharan Africa: An Analysis of the African Plant Checklist and Database'. *Taxon* 56: 201–08.
19 A. P. Davis T. W. Gole, S. Baena *et al.* 2012. 'The Impact of Climate Change on Indigenous Arabica Coffee (Coffea arabica): Predicting Future Trends and Identifying Priorities'. *PLoS ONE* 7: e47981. https://doi.org/10.1371/journal.pone.0047981.
20 K. C. Chinsembu. 2015. 'Plants as Antimalarial Agents in Sub-Saharan Africa'. *Acta Tropica* 152: 32–48. https://doi.org/10.1016/j.actatropica.2015.08.009.
21 J. M. MacKenzie. 1997. *The Empire of Nature: Hunting, Conservation and British Imperialism* (Manchester: Manchester University Press).
22 A. G. Rebelo. 1992. 'Red Data Book Species in the Cape Floristic Region: Threats, Priorities and Target Species'. *Transactions of the Royal Society of South Africa* 48: 55–86. https://doi.org/10.1080/00359199209520256.
23 D. Brugiere and R. Kormos. 2009. 'Review of the Protected Area Network in Guinea, West Africa, and Recommendations for New Sites for Biodiversity Conservation'. *Biodiversity and Conservation* 18: 847–68. https://doi.org/10.1007/s10531-008-9508-z.
24 M. Cioc. 2009. *The Game of Conservation: International Treaties to Protect the World's Migratory Animals* (Columbus: Ohio University Press).

25 D. K. Prendergast and W. M. Adams. 2003. 'Colonial Wildlife Conservation and the Origins of the Society for the Preservation of the Wild Fauna of the Empire (1903–1914)'. *Oryx* 37: 251–60. https://doi.org/10.1017/S0030605303000425.
26 African Parks. 2016. *African Parks Annual Report 2015: Conservation at Scale* (Johannesburg: African Parks). www.africanparks.org/sites/default/files/uploads/resources/2017-05/APN_AnnualReport_2015.pdf.
27 R. Cooney, D. Roe, H. Dublin *et al.* 2017. 'From Poachers to Protectors: Engaging Local Communities in Solutions to Illegal Wildlife Trade'. *Conservation Letters* 10: 368–74. https://doi.org/10.1111/conl.12294.
28 S. Canney and N. Ganamé. 2015. 'The Mali Elephant Project, Mali'. In: *Conservation, Crime and Communities: Case Studies of Efforts to Engage Local Communities in Tackling Illegal Wildlife Trade*, ed. D. Roe (London: IIED). http://pubs.iied.org/14648IIED/.
29 J. Kalpers, E. A. Williamson, M. M. Robbins *et al.* 2003. 'Gorillas in the Crossfire: Population Dynamics of the Virunga Mountain Gorillas over the Past Three Decades'. *Oryx* 37: 326–37. https://doi.org/10.1017/S0030605303000589.
30 D. Roe, F. Nelson and C. Sandbrook. 2009. 'Community Management of Natural Resources in Africa: Impacts, Experiences and Future Directions'. *Natural Resource Issues* 18 (London: IIED). http://pubs.iied.org/pdfs/17503IIED.pdf.
31 J. King, A. Wandera, M. I. Sheikh *et al.* 2016. Status of Hirola in Ishaqbini Community Conservancy (Masalani: Ishaqbini Hirola Community Conservancy; Northern Rangelands Trust). https://nrt-kenya.squarespace.com/s/Hirola-sanctuary-status-report-May2016-mjwy.pdf
32 S. Richter. 2011. 'What is a Lion Worth? Valuing Africa's Tourism Potential'. *Tourism and Leisure*. October.
33 J. Benson. 2018. *Africa's Blue Economy*. One Earth Future Fact Sheet, USA. 5 pp.
34 J. A. Oldekop, G. Holmes, W. E. Harris *et al.* 2016. 'A Global Assessment of the Social and Conservation Outcomes of Protected Areas'. *Conservation Biology* 30: 133–41. https://doi.org/10.1111/cobi.12568.
35 S. Hauenstein, M. Kshatriya, J. Blanc *et al.* 2019. 'African Elephant Poaching Rates Correlate with Local Poverty, National Corruption and Global Ivory Price'. *Nature Communications* 10: 2242. https://doi.org/10.1038/s41467-019-09993-2.
36 H. Cross. 2015. 'Displacement, Disempowerment and Corruption: Challenges at the Interface of Fisheries, Management and Conservation in the Bijagós Archipelago, Guinea-Bissau'. *Oryx* 50: 693–701. https://doi.org/10.1017/S003060531500040X.
37 J. Nackoney, G. Molinario, P. Potapov *et al.* 2014. 'Impacts of Civil Conflict on Primary Forest Habitat in Northern Democratic Republic of the Congo, 1990–2010'. *Biological Conservation* 170: 321–28. https://doi.org/10.1016/j.biocon.2013.12.033.
38 J. C. Brito, S. M. Durant, N. Pettorelli *et al.* 2018. 'Armed Conflicts and Wildlife Decline: Challenges and Recommendations for Effective Conservation Policy in the Sahara-Sahel'. *Conservation Letters* 11: e12446. https://doi.org/10.1111/conl.12446.
39 J. H. Daskin and R. M. Pringle. 2018. 'Warfare and Wildlife Declines in Africa's Protected Areas'. *Nature* 553: 328–32. https://doi.org/10.1038/nature25194.
40 A. Amano, T. Székely, B. Sandel *et al.* 2018. 'Successful Conservation of Global Waterbird Populations depends on Effective Governance'. *Nature* 553: 199–202. https://doi.org/10.1038/nature25139.
41 W. F. Laurance 2013. 'Does Research Help to Safeguard Protected Areas?' *Trends in Ecology and Evolution* 28: 261–66. https://doi.org/10.1016/j.tree.2013.01.017.
42 K. A. Wilson, N. A. Auerbach, K. Sam, et al. 2016. 'Conservation Research is Not Happening Where it Is Most Needed'. *PLoS Biology* 14: e1002413. https://doi.org/10.1371/journal.pbio.1002413.

43 P. Koohafkan, M. Salman and C. Casarotto. 2011. 'Investments in Land and Water'. *SOLAW Background Thematic Report TR17* (London: FAO). www.fao.org/fileadmin/templates/solaw/files/thematic_reports/TR_17_web.pdf.

44 C. Leisher, Robinson, N., Brown, M., Kujirakwinja, D., Schmitz, M. C., Wieland, M. and Wilkie, D. (2020). 'Ranking the Direct Threats to Biodiversity in sub-Saharan Africa'. *BioRXIV Ecology*: 1–23.

45 BirdLife International. 2019. *IBA's in Danger-Site Summary.* http://datazone.birdlife.org/site/ibaidsites.

46 http://whc.unesco.org/en/danger.

3 Fisheries and aquaculture in the western Indian Ocean states

Erika Techera, PhD and Abdi Fatah Hassan, JD

Introduction

The blue economy has gained considerable traction as a way to derive wealth from the oceans and to sustainably balance development goals, environmental protection and social equity.

The blue economy concept first emerged in international governance at the United Nations (UN) Conference on Sustainable Development in 2012,[1] and has since been championed by island states such as Seychelles,[2] and endorsed by key organizations such as the African Union (AU) and the Indian Ocean Rim Association (IORA).[3] The blue economy has no universal definition but has been described as the achievement of economic growth and development based on ocean activities, together with enhanced social and environmental outcomes.[4] Clearly, strong links exist between the blue economy and sustainable development concepts;[5] indeed, 'sustainable development in relation to oceans, is a key component of the blue economy'.[6] Therefore, adopting a blue economy agenda is highly attractive for coastal states, as it provides the opportunity to sustainably develop economies and reduce poverty while also protecting marine ecosystems and safeguarding ocean health.

As discussed in other chapters in this volume, the blue economy involves various sectors including fisheries, energy, shipping, mining and tourism. Expanded fishing and aquaculture[7] are a key feature of the blue economy because of the multiplicity of beneficial outcomes: food security, health benefits, sustainable livelihood opportunities and the economic development of renewable resources. Achieving blue economy and sustainable development goals requires the maintenance of fish stocks and the broader ocean health; yet this has proved challenging in practice.[8] Balancing sustainable harvests with protecting the environment has long been the subject of governance initiatives, with varying success. Internationally, multiple institutions are involved, including the Food and Agriculture Organization of the United Nations (FAO), and regionally relevant inter-governmental and fisheries organizations. Soft law initiatives and hard law treaties have created frameworks for sustainable fisheries governance, by setting standards, providing best practice guidance and creating forums for discussion. Added to these focused fisheries initiatives are the broader UN Sustainable Development Goals (SDGs), which seek to eradicate poverty and hunger, improve food security, nutrition, health and well-being, promote sustainable economic growth and full employment, ensure sustainable consumption and production, and conserve and sustainably use ocean and marine resources.[9] The SDGs provide valuable global targets and indicators, an element missing from the blue economy literature to date. At the domestic level, all nations have fisheries regulations and government agencies tasked with developing policies and laws, implementing legal measures and managing stocks, monitoring compliance and enforcing rules (although enforcement varies greatly from state to state). This chapter explores these governance arrangements, and analyses whether they are adequate to support sustainable development and blue economy goals focused on expanded fisheries and aquaculture sectors.

The chapter focuses on the western Indian Ocean (WIO) states, including continental East Africa (South Africa, Mozambique, Tanzania, Kenya and Somalia) and the African island states (Mauritius, Seychelles, Comoros and Madagascar). This focus is justified because fisheries have, and continue to be, important for food, livelihoods and economies in these states, and therefore good fisheries governance is critical. Furthermore, as the ocean areas are interconnected and fish stocks shared, and also because these states are linked through inter-governmental organizations and agreements, it is valuable to explore this subregion holistically. In addition, there are common risks, given the significant uptake of ocean economy policies in this region. If all WIO states pursue blue growth goals, and expand their fisheries and aquaculture sectors, increasing pressure will be placed on already depleted wild fishery resources and associated species, as well as on the ocean environment given the risks associated with aquaculture.[10] Therefore, identifying ways to enhance the governance of fisheries and aquaculture, which provides a robust foundation for blue growth, is vital. A governance lens has been adopted because the risks posed by weak and ineffective frameworks, as well as the challenge of policy coherence and coordination, have already been identified as key issues for sustainable fisheries and aquaculture in the region.[11]

This chapter commences with an examination of how WIO states have engaged with the concepts of sustainable development and the blue economy. In particular, policies featuring an expansion of the fisheries and aquaculture sectors are considered. The next section assesses the relevant international and regional hard and soft law fisheries and aquaculture obligations to which states have committed. This is followed by an analysis of the extent to which national fisheries and aquaculture governance incorporates the concepts of sustainable development and the blue economy, and an assessment of whether these frameworks are likely to be effective in achieving triple bottom line goals. The final section highlights key considerations for the future.

The blue economy, SDGs and fisheries in Africa

Fish and fisheries are of crucial importance to populations across the WIO region.[12] In 2018 the total live tonnage of fish caught in marine areas in the region was 5.5 million metric tons, representing 45% of the total catch for the Indian Ocean as a whole.[13] On a per capita basis, however, fish consumption across East Africa is less than one-half that of West Africa and other parts of the continent.[14] Despite the nutritional benefits and the relative abundance of marine resources, the FAO has attributed this relatively lower consumption rate to factors including populations 'increasing at a higher rate than food fish supply; stagnation of fish production because of pressure on capture fisheries resources; and a poorly developed aquaculture sector'.[15] This context necessitates an examination of blue economy policies in this region as a means of balancing food security, economic development and sustainable use of marine resources.

The WIO states' blue economy strategies vary significantly for a number of reasons. On the one hand, Kenya, Seychelles, South Africa and Mauritius have made significant efforts to develop blue economy strategies that range from improving port facilities, supporting artisanal fishermen, developing aquaculture sectors and maintaining maritime security. In contrast, Somalia and other less developed East African states have struggled to formulate blue economy strategies. The AU has said that in order to tackle issues affecting the continent as a whole – waste generation, climate change, energy and economic growth – a blue economy agenda is needed to improve 'Africans' well-being while significantly reducing marine environmental risks'.[17] So, in general, WIO states have broadly engaged with the blue economy, albeit in diverse ways.[18] In terms of fisheries and aquaculture development, capacity building is key,[19] and less-developed states are likely to see greater relative socio-economic benefits from improved fisheries governance. In states such as Mauritius, fishing accounts for only 1% of national gross domestic product,

Table 3.1 Relative importance of the fisheries sector to WIO states[16]

	Fisheries output (% of national gross domestic product)	*Employment in the fisheries sector (% of population)*
Comoros	Unknown	11,400 (1.5%)
Kenya	0.54%	2,000,000 (4.4%)
Madagascar	7%	100,000 (0.44%)
Mauritius	1%	29,055 (2.29%)
Mozambique	10.3%	202,000 (0.79%)
Seychelles	1.2%	1,800 (2%)
Somalia	1%	Unknown
South Africa	>1%	16,800 (0.02%)
Tanzania	1.4%	4, 000,000 (7.14%)

although there is positive growth in this sector, and the implementation of a blue economy agenda would encourage further development.[20] Mauritius has made positive policy moves to realize the potential benefits of a blue economy agenda by refocusing its Ministry of Blue Economy, Marine Resources, Fisheries and Shipping to expand contributions to the wider economy.[21] For its efforts, Mauritius is increasingly viewed as a leader in this area.[22] Kenya took similar steps to rename its fisheries agency as the State Department for Fisheries, Aquaculture and the Blue Economy,[23] although its efforts have been largely restricted to infrastructure projects and maritime security.[24] Legislatively, few states have embedded blue economy principles in hard law,[25] and this is seen as a necessary step in achieving blue economy goals.[26]

Across the WIO region, small-scale fishing makes up a substantial proportion of the ocean economy.[39] Conversely, aquaculture remains an underdeveloped sector for most states, with Tanzania and Madagascar being the largest producers.[40] Evidence demonstrates that fishing communities possess the necessary skills to provide food and livelihood security for themselves, and that small-scale fishing poses fewer risks to the marine environment when compared to large-scale commercial activities.[41] Supporting local fishing communities, through policies that incorporate good fisheries governance, therefore has the potential to help WIO states to meet SDG 2 (food security) and SDG 14 (safeguarding life below water), as well as SDG 1 (ending poverty) and SDG 8 (promoting sustainable economic growth and full employment). Currently, there are policy gaps as some states are failing to clearly articulate their blue economy strategies (Comoros, Somalia) while others are mostly focusing on their economic goals (Kenya, Madagascar) that prevent local WIO fishing communities from deriving sustainable economic growth through fishing, which in turn poses governance challenges for states. Opportunities remain, of course, for these states to increase the economic output of local communities while protecting the marine environment through the development and implementation of sound blue economy strategies.

Legal governance of fisheries and aquaculture in the WIO

Achieving sustainable fisheries governance is challenging, because fish are mobile, stocks hard to quantify and fishing activities challenging to monitor. Significant efforts have been made to advance sustainable fisheries, at all governance levels, each of which is explored below.

Globally, the FAO is the principal UN institution with a mandate to address food security, including playing a leading role in international fisheries policy through the work of its Committee on Fisheries.[42] The FAO has established regional fisheries bodies and set standards for fisheries and aquaculture through the adoption of codes of conduct and plans of action, and has provided critical global catch and production data.[43] There is, however, no overarching international fisheries treaty. The UN Convention on

Table 3.2 Blue economy policies in WIO states[27]

	National body responsible for the blue economy	*Policy, strategy and goals*
Comoros	General Directorate of Fishery Resources (Direction Générale des Ressources Halieutiques)	No specific blue economy strategies – relevant policies are largely drafted in terms of 'sustainability' and focused on improving regional collaboration, data sharing, illegal fishing and foreign direct investment.[28]
Kenya	State Department for Fisheries, Aquaculture and the Blue Economy	Much of the department's blue economy policy is aimed at increasing investment in port facilities and preventing maritime crime.[29] There is no specific blue economy policy directed towards fisheries, and the only marine environmental goals that are addressed is the protection of coastal mangroves and seagrass.
Madagascar	Ministry of Environment, Ecology and Forests	Although no clear policy has been outlined, the Malagasy government's blue economy agenda has focused on infrastructure and tourism.[30] Elsewhere, efforts directed towards the environment, such as the establishment of community-led marine protected areas and sustainable fishing practices, have relied on the support of non-governmental organizations and regional actors.[31]
Mauritius	Ministry of Blue Economy, Marine Resources, Fisheries and Shipping	A specific policy addresses increasing aquaculture production with a focus on sustainability.[32] In terms of supporting the marine environment for fisheries, the approach taken by the ministry has been to improve governance through the phasing out of certain fishing practices (nets) and the creation of licencing/permit systems (pilot aquaculture programme).[33]
Mozambique	Ministry of the Sea, Inland Waters and Fisheries	Created in 2015 to further Mozambique's blue economy agenda, much of the ministry's work has relied on the support of international actors including the World Bank and the South West Indian Ocean Fisheries Commission. The ministry's areas of focus are research and climate change; ecosystem rehabilitation; infrastructure; sustainable economic growth; spatial planning; and addressing illegal, unreported and unregulated fishing.[34]

	National body responsible for the blue economy	*Policy, strategy and goals*
Seychelles	Department of the Blue Economy	A central government policy underpins the department's blue economy agenda comprising four key pillars: (1) economic diversification and resilience; (2) shared prosperity; (3) food security; and (4) integrity of habitats and ecosystem services.[35] This policy has a specific focus on diversifying and increasing the output of fisheries, although this is counterbalanced by the four pillars mentioned above which promote sustainable and equitable development.
Somalia	Ministry of Fisheries and Marine Resources	No specific policy document articulates a blue economy agenda, although the 'conservation of coastal fisheries ecosystem' and the collection of data to support 'sustainable harvesting of fish in Somalia territorial waters' are among the goals of its fisheries.[36]
South Africa	Ministry of Environment, Forestry and Fisheries	No blue economy policy specific to fisheries, although, every two years, a review of the fisheries sector has been conducted, and in comparison to other WIO states it has relatively more comprehensive data on fish stock levels.[37]
Tanzania	Ministry of Livestock and Fisheries	A recent ministerial policy instrument does not make reference to the blue economy.[38] Nevertheless, the current policy recognizes that fish stocks will be depleted if they are not sustainably managed. Notably, the policy addresses the need for capacity building, an ecosystem-based approach to governance, and the need to develop an aquaculture sector.

the Law of the Sea (UNCLOS) establishes maritime zones including the territorial sea, exclusive economic zones (EEZs) and the high seas.[44] Relevantly, UNCLOS confirms that states have sovereignty over the territorial sea, and can exploit living marine resources subject only to a general obligation to protect and preserve the marine environment.[45] In the EEZs, the coastal states have the sovereign right to explore and exploit living marine resources, but must determine the total allowable catch for a given species by taking into account the best scientific evidence, involving an assessment of the maximum sustainable yield and adopting conservation and management measures to avoid over-exploitation.[46] If a state cannot fully exploit the marine resources of its EEZ then it must offer the right to do so to other nations, again with the goal of achieving the optimum utilization of the EEZ's living resources.[47] Under UNCLOS, the high seas are open to all nations to fish, subject to obligations to adopt measures for the conservation and management of living resources.[48] Again, the objective of producing the maximum sustainable yield is a key element. Given the widespread ratification of UNCLOS, its framework provides a solid foundation and impetus for sustainable fisheries governance. Relevant laws have been adopted at the domestic level in order to regulate national fisheries through regional fishery management organizations (discussed further below). While UNCLOS makes no specific reference to sustainable development or the blue economy, the provisions noted above do seek to avoid unsustainable harvests and to achieve a balance between use and conservation. The Convention does not refer to aquaculture but does refer to structures and installations which may be relevant to these activities.[49] Nevertheless, broader obligations to protect and preserve the marine environment would apply.

The only other relevant globally binding instrument is the UN Fish Stocks Agreement (UNFSA) which focuses on highly migratory and straddling stocks in high seas areas.[50] Drafted more than a decade after UNCLOS, the UNFSA incorporates the objective of 'conservation and sustainable use' of fish stocks.[51] Yet it echoes UNCLOS in requiring states to 'adopt measures to ensure long-term sustainability of straddling fish stocks' aimed at optimum utilization to produce maximum sustainable yields.[52] As the UNFSA focuses on wild stocks, it does not address aquaculture. The Convention on Biological Diversity (CBD) is also relevant, as it seeks to balance conservation with sustainable use of biodiversity – including fish – and the equitable sharing of benefits.[53] Meanwhile, the Aichi Targets have been adopted under the CBD, several of which are relevant:

> Target 6: By 2020 all fish … are managed and harvested sustainably [and] legally … so that overfishing is avoided … fisheries have no significant adverse impacts on threatened species and vulnerable ecosystems and the impacts of fisheries on stocks, species and ecosystems are within safe ecological limits.
> Target 7: By 2020 areas under agriculture, aquaculture and forestry are managed sustainably, ensuring conservation of biodiversity.[54]

While not legally binding upon states, these targets are highly influential, and combined with the SDGs referred to above, provide clear goals for states. Although a binding instrument is lacking, the FAO has adopted soft laws that provide guidance, principles and standards on best practice fishing and aquaculture governance. Most relevantly, the Code of Conduct on Responsible Fisheries (CoC-RF) explicitly refers to 'food security, poverty alleviation and sustainable development' in its general principles.[55] Sustainable use and management are referred to repeatedly throughout,[56] and social and cultural factors and impacts are recognized as important elements in responsible management and resource allocation.[57] The CoC-RF calls upon states to implement policy, legal and institutional frameworks to balance conservation and long-term sustainable use of fisheries resources and/or aquaculture development, and to ensure that operations are carried out responsibly.[58] Nevertheless, optimum utilization is still promoted.[59]

These legal instruments provide a basis for balancing the use of marine fisheries resources and the conservation of species and environments. However, accurately determining maximum yields and ensuring optimum utilization in the context of advancing blue economy goals is fraught with difficulty. Very little binding international law addresses aquaculture, and so the voluntary CoC-RF is the principal global instrument mandated with setting standards for developing this sector. Another key issue is the absence of any reference to sociocultural concerns in the treaties. This aspect is an important element of the sustainable development triple bottom line and is also a feature of the blue economy. Again, it is only the CoC-RF that places any emphasis on poverty eradication, social impact and cultural factors. Therefore, international regimes, in isolation, provide an incomplete foundation to guide the enhancement of fisheries and aquaculture governance to achieve blue economy goals.

A number of regional fisheries bodies have also been established that focus on wild fish stocks and fishing in areas beyond national jurisdiction, but these fisheries bodies do not address fishing in the territorial sea and EEZs so are not considered here. IORA is the only relevant regional inter-governmental organization comprising all WIO states. IORA has identified fisheries management as a priority area, and the blue economy as a key focus.[61] Despite several significant meetings, and the IORA Action Plan 2017–2021, no binding regional agreements have been adopted. It remains to be seen which binding IORA initiatives will emerge. Other relevant regional bodies include the Common Market for Eastern and Southern Africa (COMESA) which

Table 3.3 WIO state membership of key bodies and ratification of treaties[60]

State	*International law*			*International organization*				
	UNCLOS	UNFSA	CBD	FAO	COMESA	EAC	IOC	IORA
Comoros	1994	–	1994	1977	X		X	X
Kenya	1989	2004	1994	1964	X	X		X
Madagascar	2001	–	1996	1961	X		X	X
Mauritius	1994	1997	1992	1968	X		X	X
Mozambique	1997	2008	1995	1977	X			X
Seychelles	1991	1998	1993	1977	X		X	X
Somalia	1989	–	2009	1960				X
South Africa	1997	2003	1996	1993				X
Tanzania	1985	–	1996	1962	X	X		X

has a Fisheries Program with a priority area focused on fisheries management and governance including interventions to avoid unsustainable exploitation.[62] The Indian Ocean Commission (IOC), whose membership includes the African island states, has adopted a Regional Fisheries and Aquaculture Strategy (2015–2025) highlighting blue growth potential and noting weaknesses in governance.[63]

Turning to the national level, the analysis below explores how the WIO states have implemented their international obligations and included national sustainable development and blue economy goals. Most of the states have multiple policies that have an impact on the fisheries and aquaculture sectors, as set out in Table 3.4. The WIO states' fisheries policies do not commonly refer to the blue economy,[64] although some national biodiversity strategies make reference to both sustainable development and blue growth.[65] Seychelles (see Chapter 9 in this volume) – which has championed the blue economy concept – is the only nation whose fisheries policy incorporates the concept of the blue economy and the SDGs, and emphasizes the economic, sociocultural and environmental benefits to be gained therefrom.[66]

Despite the absence of explicit references to the blue economy, most WIO states do refer to sustainable development.[68] Furthermore, most states also make specific reference to balancing conservation and the sustainable use of fisheries resources, as well as to the objective of optimum utilization.[69] This demonstrates implementation of key UNCLOS and other international obligations.[70] Most states have also sought to balance conservation and sustainable use of fisheries resources with social and cultural goals.[71] South Africa, for example, aims to provide an 'effective response to poverty and food insecurity by maximising the use and management of natural resources to create vibrant, equitable and sustainable rural communities',[72] and refers to 'job creation, food security, rural development', in addition to economic and environmental outcomes, as relevant considerations when allocating fishing rights (see Chapter 8 in this volume).[73] This acknowledges the sociocultural aspects inherent in the blue economy as well as sustainable development. Some states highlight the significant small-scale fisheries sector.[74] South Africa has also developed a specific policy on small-scale fisheries with objectives referring explicitly to sustainable development, including the interdependencies of sociocultural, economic and ecological dimensions, the need to ensure that benefits flow to the relevant communities and the value of cooperative governance 'to promote poverty alleviation, food security, sustainable livelihoods, fair and safe labour practices, and local economic development'.[75] A final feature of the fisheries policies are references to opportunities to develop niche products, processing activities and value-added services.[76] Worryingly, subsidies are mentioned as the means through which to achieve these goals, which runs counter to SDG 14.6 and Aichi Biodiversity Target 3, which support subsidy phase-outs.[77]

Specific aquaculture policies are less common among these nations.[78] Nevertheless, Kenya's Fisheries Sector Policy and Strategy includes a section

Table 3.4 Relevant WIO fisheries and aquaculture policies and legislation[67]

State	*Policies*	*Laws*
Comoros	National Strategy and Action Plan for Biodiversity (2016) Comorian Fisheries Development Strategy (2007)	Law on Comorian Fishing and Aquaculture, 2007 Fisheries and Aquaculture Code
Kenya	National Ocean and Fisheries Policy (2008) Kenya Tuna Fisheries Development and Management Strategy (2013)	Fisheries Management and Development Act 2016
Madagascar	National Strategy for Good Governance of Maritime Fisheries in Madagascar (2012) National Biodiversity Strategy and Action Plans (2015) National Development Plan 2015–2019	Decree no. 2016–1492 relating to the general reorganization of sea fishing activities (2016) Decree no. 2016–1493 regulating aquaculture activities Law no. 2015–053 on the Fisheries and Aquaculture Code (2015)
Mauritius	National Biodiversity Strategy and Action Plan 2017–2025	Fisheries and Marine Resources Act 2007 Fisheries and Marine Resources (Fish Farming) Regulations 2014 Environment Protection Act 2002
Mozambique	Fisheries Plan 2010–2019 National Strategy and Action Plan of Biological Diversity of Mozambique (2015–2035)	Fisheries Law 2013
Seychelles	Fisheries Sector Policy and Strategy 2019 National Biodiversity Strategy and Action Plan (NBSAP) 2015–2020 Ministry of Fisheries and Agriculture Strategic Plan 2018–2020	Fisheries Act 2014 Fisheries (Shark Finning) Regulations 2006 Fisheries Regulations 1987 as amended
Somalia	National Development Plan 2017–2019 National Biodiversity Strategy and Action Plan 2015–2020 The Somali Compact (2014)	Fisheries Law 1985
South Africa	Policy for the Small-Scale Fisheries Sector in South Africa (2012) General Policy on the Allocation and Management of Fishing Rights (2013) South Africa's 2nd National Biodiversity Strategy and Action Plan (2015)	Marine Living Resources Act 1998 Sea Fishery Act 1988

State	*Policies*	*Laws*
Tanzania	National Fisheries Policy (2015) National Biodiversity Strategy and Action Plan 2015–2020 National Five-Year Development Plan 2016–2021	Fisheries Act 2003

on aquaculture development which refers to commercial approaches, the use of sustainable technologies and social goals such as community participation including women and youth.[79] Tanzania's National Five-Year Development Plan also refers to aquaculture including the future establishment of a national aquaculture development programme, but with a strong emphasis on economic outcomes.[80] Further detail is provided in Tanzania's National Fisheries Policy where aquaculture opportunities are well recognized and viewed as a way to contribute to poverty reduction and enhanced food security.[81] Again, Seychelles' Fisheries Sector Policy and Strategy is more advanced, incorporating a major policy goal to 'promote an environmentally responsible aquaculture industry that contributes towards food security and the creation of wealth in the Seychelles'.[82] Good governance is an explicit strategy, alongside a sustainable environment, investment, inclusive participation and research.[83]

All the WIO states have relevant fisheries laws, as set out on Table 3.4 above. In general, this legislation establishes a governance institution,[84] a licensing, permitting and registration regime,[85] development, planning and conservation measures,[86] compliance[87] and enforcement provisions.[88] These legal arrangements provide a basis for sound fisheries management. Some jurisdictions cover additional issues such as marketing and trade regulations including those relating to marketing authorities, import and export arrangements,[89] quality and safety,[90] differential arrangements for artisanal, commercial and foreign fishers,[91] and establish development funds.[92] Relevantly, the Kenyan legislation includes objectives that incorporate sustainability with various references to triple bottom line goals, although none refer to the blue economy.[93] These objectives are supported by principles that also refer to the long-term sustainable use of fishery resources, sound management, optimum utilization, economic growth, conservation, participation and equity.[94] In Tanzania, the act refers to the 'development and sustainable use' of fish stocks and aquatic resources, but detailed principles are not articulated.[95] In South Africa, the Sea Fishery Act does not include objectives or principles, but does authorize the minister responsible for fisheries to make policies for the 'conservation and optimal utilization' of living marine resources to achieve 'protection and sustained utilization of the sea' with regard to 'economic, social and cultural values'.[96] However, no further reference to sustainability is made. The more recent Marine Living Resources Act includes objectives and principles including the 'need to achieve optimum utilization and ecologically sustainable development of marine living resources'.[97] The need to achieve 'economic growth, human resource development ... employment creation' is cited, and to 'restructure the fishing industry to address historical imbalances and to achieve equity', as well as broad participation in decision-making processes.[98] The Mauritian and Seychellois statutes do not include objectives or principles, and make no reference to sustainable development.

Aquaculture laws are less developed in WIO states, yet some jurisdictions have relevant governance arrangements. Kenya includes aquaculture within its Fisheries Law covering development planning, promotion and permitting, the role of local government, safeguards for traditional fish farming, waste management and environmental protection, inspection and enforcement provisions.[99] Tanzania also includes aquaculture within the Fisheries Act, and requires development to be ecologically sustainable with resources rationally used.[100] Interestingly, reference is made to ensuring that aquaculture development does not interfere with 'livelihood, culture and traditions of local communities and their access to fishing ground', but the way in which fish farming can achieve sociocultural objectives is not mentioned.[101] In South Africa, the Marine Living Resources Act refers to developing mariculture, including the power to make regulations for the orderly development of the sector, but established no detailed regulatory framework.[102] Seychelles' Fisheries Act sets out a process for aquaculture permitting and operations, but nothing further.[103] Mauritius has established fish farming zones, permitting and requirements to notify of a disease outbreak under its Fisheries and Marine Resources Act,[104] and has developed fish farming regulations. The pollution risks of aquaculture are also recognized, as fish farming is a listed activity requiring environmental impact assessment under the Environmental Protection Act.[105]

In summary, there are fisheries and aquaculture governance arrangements in WIO states, but these are neither consistent nor comprehensive. They appear to have developed organically as the need arose, and generally are not well aligned with sustainable development and blue economy policy goals.

Conclusion

The blue economy is a relatively new concept and most of the laws and policies explored above pre-date its emergence. Blue economy policies with references to fisheries do exist in some states, but more commonly fisheries policies refer to sustainability. Of note is Somalia's National Development Plan, which makes detailed reference to the SDGs, mapping sector policies against SDG indicators.[106] That plan covers all sectors and provides a holistic analysis of where policy goals align with the SDGs, together with baselines data and targets, which serve to highlight gaps and challenges. Although its implementation has been hindered by general governance issues, it serves as a useful model for other WIO states where to date approaches have been piecemeal.[107]

The influence of global sustainability initiatives can clearly be seen in efforts to achieve the CBD Aichi Targets,[108] but so too can the influence of UNCLOS and the concept of optimum utilization of resources.[109] As noted above, there are risks that economic goals and optimum utilization will override sustainability, conservation and sociocultural concerns. This highlights the need for integration of governance arrangements to avoid fragmentation and competing principles and priorities. Seychelles has the most developed policy landscape in this context, with its Fisheries Sector Policy and Strategy committing the government to '[d]evelop memoranda of understanding to govern critical interagency relationships and integrate the fisheries and aquaculture sector in the broader context of Blue Economy'.[110] This approach could – and should – be utilized by other nations seeking to integrate blue economy, sustainable development and fisheries/aquaculture goals.

To achieve fisheries and aquaculture blue economy goals, robust and effective legal frameworks are essential. More recent legislation is better equipped to meet the challenges posed by the blue economy. Kenyan fisheries law, for example, is sophisticated and responsive to current and future needs. Good governance also involves effective institutions and administrative frameworks. As noted above, Kenya, Seychelles and Mauritius have reformed government agencies to refer explicitly to the blue economy. There are lessons to be learnt by other states from these approaches.

In many states, law and policy gaps can be seen. For example, the potential of aquaculture is frequently highlighted but has yet to fully develop, and states have been slow to address law and policy gaps.[111] Tailored policies, supported by administrative and legal frameworks, are needed to stimulate the sector. While rudimentary permitting regimes have sufficed to date, these are inadequate for the future if aquaculture is to be significantly scaled up, involving multiple sites and species with concomitant environmental and socio-economic risks. Nevertheless, laws must support sustainable aquaculture while not hampering it with over-regulation.[112] Mauritius has some of the most detailed aquaculture provisions for fish farming zones and disease protection, but effective governance must also include sociocultural objectives, monitoring and water quality provisions, together with arrangements for participation and markets for products.

The blue economy has been embraced by WIO states, with fisheries and aquaculture emerging as key sectors for development. These opportunities must be balanced with adequate protection of resources and environments. While all the states explored in this chapter have some governance arrangements relevant to fisheries-related blue economy goals, there is scope for improvement. Domestic imperatives have driven a diverse range of regional approaches, and there is value in comparing arrangements in each nation. Thus, sub-Saharan African states can learn from each other to enhance fisheries governance and achieve their ambitious blue economy goals.

Notes

1 Outcome document of the UN Conference on Sustainable Development, 'The Future We Want', A/RES/66/288 (2012).
2 See Chapter 9 in this volume and *Seychelles' Blue Economy, Strategic Policy Framework and Roadmap: Charting the Future (2018–2030)*, accessed 14 July 2020, https://seymsp.com/wp-content/uploads/2018/05/CommonwealthSecretariat-12pp-RoadMap-Brochure.pdf.
3 'Blue Economy', Indian Ocean Rim Association, accessed 14 July 2020, www.iora.int/en/priorities-focus-areas/blue-economy.
4 Michelle Voyer, Clive Schofield, Kamal Azmi, Robin Warner, Alistair McIlgorm and Genevieve Quirk, 'Maritime Security and the Blue Economy: Intersections and Interdependencies in the Indian Ocean', *Journal of the Indian Ocean Region* 14(1) (2018): 28–48, 29.
5 Defined as 'development that meets the needs of the present without compromising the ability of future generations to meet their own needs': World Commission on Environment and Development, *Our Common Future, an Annex to UN Doc A/42/427* (Oxford: Oxford University Press, 1987). Commonly, sustainable development is conceived of as requiring a balance between economic, environmental and sociocultural concerns – achieving triple bottom line goals.
6 World Bank and UN Department of Economic and Social Affairs, *The Potential of the Blue Economy: Increasing Long-Term Benefits of the Sustainable Use of Marine Resources for Small Island Developing States and Coastal Least Developed Countries* (Washington, DC: World Bank, 2017), 4.
7 For the purposes of this chapter, 'aquaculture' is used to refer to freshwater and marine fish farming.
8 For the purposes of this chapter, 'governance' is defined as including institutions, rules and processes. See UNEP, *Environmental Governance*, (Nairobi: UNEP, 2009), www.unep.org/pdf/brochures/EnvironmentalGovernance.pdf. It encompasses policies, laws and regulations, as well as the actors that develop and implement them, and the decision-making processes involved.
9 UN SDGs, accessed 21 July 2020, https://sdgs.un.org/goals.
10 Jana Roderburg, 'Marine Aquaculture: Impacts and International Regulation', *A&NZ Mar LJ* 25 (2011): 161.
11 AUC-NEPAD, 'The Policy Framework and Reform Strategy for Fisheries and Aquaculture in Africa' (2014), https://au.int/web/sites/default/files/documents/30266-doc-au-ibar_-_fisheries_policy_framework_and_reform_strategy.pdf.
12 Rudy van der Elst, 'Fish, Fishers and Fisheries of the Western Indian Ocean: Their Diversity and Status. A Preliminary Assessment', *Philosophical Transactions: Mathematical, Physical and Engineering Sciences* 363(1826) (2005): 263, 264.
13 Food and Agriculture Organization of the United Nations (FAO), *The State of World Fisheries and Aquaculture*, (Rome: FAO, 2020), 16.
14 Ibid., 70.
15 Ibid., 71.
16 Economic and employment data taken from FAO Fishery and Aquaculture Country Profiles (www.fao.org/fishery/countryprofiles/search/en) and the Indian Ocean Commission (IOC) SmartFish Programme (www.fao.org/). Population date taken from the World Bank database (https://data.worldbank.org/indicator/SP.POP.TOTL).
17 African Union, *2050 Africa's Integrated Maritime Strategy (2050 AIM Strategy)* (Addis Ababa: AU), 4.
18 Adam Moolna and Benjamin S. Thompson, 'The Blue Economy Approach for Sustainability in Seychelles & East Africa' (Discussion Paper, Institute for Sustainable Futures, Keele University, November 2018), 1.

19 J. C. Seijo and S. Salas, 'The Role of Capacity Building for Improving Governance of Fisheries and Conservation of Marine Ecosystems', in Serge M. Garcia, Jake Rice and Anthony Charles (eds), *Governance of Marine Fisheries and Biodiversity Conservation: Interaction and Co-Evolution* (Hoboken, NJ: John Wiley & Sons, 2014), 374–82.
20 FAO, Fishery Country Profile, *The Republic of Mauritius*, www.fao.org/fi/oldsite/FCP/en/MUS/profile.htm.
21 See ministry website: http://blueconomy.govmu.org/English/Pages/default.aspx.
22 Sonali Mittra, 'Blue Economy: Beyond an Economic Proposition' (Observer Research Foundation, Issue Brief no. 173, March 2017), 3.
23 See agency website: www.kilimo.go.ke/management/state-department-of-livestock-2/.
24 Alex Benkenstein, 'Prospects for the Kenyan Blue Economy' (South African Institute of International Affairs, *Policy Insights*, no. 62, July 2018): 3.
25 Erika Techera, 'Indian Ocean Fisheries Regulation: Exploring Participatory Approaches to Support Small-Scale Fisheries in Six States', *Journal of the Indian Ocean Region* 16(1) (2020): 33–42.
26 Moolna and Thompson, 1–3.
27 Data compiled from several UN websites: fishing statistics, www.fao.org; agendas and policies, www.unenvironment.org/ and www.uneca.org/; state environmental and ministerial websites; and NGO and intergovernmental bodies' websites including www.worldbank.org/, www.iora.int/en and https://au.int/.
28 FAO, *Smart Fish, Fisheries in the ESA-IO Region: Profile and Trends – Country Review 2014: Comoros* (Rome: FAO, 2014), 20–21.
29 Indian Ocean Rim Association, *Sustainable Development in the Indian Ocean Rim*, www.iora.int/media/24002/31-07-2018-kenya-iod-sustainable-dev-ior-min.pdf.
30 UN Economic Commission for Africa, *Africa's Blue Economy: A Policy Handbook*, 2016, www.uneca.org/sites/default/files/PublicationFiles/blue-eco-policy-handbook_eng_1nov.pdf.
31 UN Environment Program, *Blue Economy: Sharing Success Stories to Inspire Change*, 2015, https://wedocs.unep.org/bitstream/handle/20.500.11822/9844/-Blue_economy_sharing_success_stories_to_inspire_change-2015blue_economy_sharing_success_stories.pdf.pdf?sequence=3&isAllowed=y.
32 Government of Mauritius, Ministry of Agro-Industry and Fisheries, *Potential for Sustainable Aquaculture Development in Mauritius*, http://blueconomy.govmu.org/English/Publication/Documents/Potential%20for%20Sustainable%20Aquaculture%20in%20Mauritius.pdf.
33 Government of Mauritius, Ministry of Blue Economy, Marine Resources, Fisheries and Shipping, http://blueconomy.govmu.org/English/Documents/Annual%20Report%20on%20Performance%2016-17%20-%20Min%20of%20Ocean%20Economy.pdf.
34 World Bank, *Communities Livelihoods Fisheries: Governance, Growth & the Blue Economy in Mozambique*, 2019, http://documents1.worldbank.org/curated/pt/194861558072729691/pdf/Communities-Livelihoods-Fisheries-Governance-Growth-and-the-Blue-Economy-in-Mozambique.pdf.
35 *Seychelles Blue Economy Strategic Framework and Roadmap*, https://seymsp.com/wp-content/uploads/2018/05/CommonwealthSecretariat-12pp-RoadMap-Brochure.pdf.
36 Government of Somalia, Ministry of Fisheries and Marine Resources, https://mfmr.gov.so/en/ministry/.
37 Government of South Africa, Status of the South African marine fishery resources, www.environment.gov.za/sites/default/files/reports/statusofsouthafrican_marinefisheryresources2016.pdf.

38 Government of Tanzania, Ministry of Livestock and Fisheries Development, *National Fisheries Policy of 2015.*

39 FAO, *Towards the Implementation of the SSF Guidelines in Eastern Africa* (Rome: FAO, 2015), 9.

40 FAO, *The State of World Fisheries and Aquaculture*, (Rome: FAO, 2020), 26–28.

41 Jeppe Kolding, Christophe Béné and Maarten Bavinck, 'Small-Scale Fisheries: Importance, Vulnerability, and Deficient Knowledge', in Serge M. Garcia, Jake Rice and Anthony Charles (eds), *Governance of Marine Fisheries and Biodiversity Conservation: Interaction and Co-Evolution* (Hoboken, NJ: John Wiley & Sons, 2014), 317–20.

42 FAO, 'Fisheries', accessed 14 July 2020, www.fao.org/fisheries/en/.

43 FAO, 'State of World Fisheries and Aquaculture'.

44 UN Convention on the Law of the Sea (adopted 10 December 1982, entered into force 16 November 1994), 1833 UNTS 3.

45 UNCLOS, Art. 193. The territorial sea extends from the baseline to 12 nautical miles Art. 3.

46 UNCLOS, Art. 61.

47 UNCLOS, Art. 62.

48 UNCLOS, Arts 116–19.

49 UNCLOS, Arts 56 and 60.

50 Agreement for the Implementation of the Provisions of the UN Convention on the Law of the Sea of 10 December 1982 Relating to the Conservation and Management of Straddling Fish Stocks and Highly Migratory Fish Stocks, (adopted 4 August 1995, in force 11 December 2001), 34 ILM 1542.

51 UNFSA, Art. 2.

52 UNFSA, Art. 5.

53 CBD, Art. 1.

54 CBD, 'Aichi Biodiversity Targets', accessed 17 July 2020, www.cbd.int/sp/targets/. Target 3 is also relevant as it calls for the phasing out of subsidies.

55 FAO, *Code of Conduct for Responsible Fisheries* (Rome: FAO, 1995), Art. 6.

56 For example, see multiple references in Art. 7 on fisheries management and Art. 9 on aquaculture development.

57 See CoC-RF, Arts 2(a), 6.4, 7.4.5, 7.6.7, 8.4.8, 9.1.5, 10.2 and 11.1.

58 Ibid., Arts 7, 8 and 9.

59 Ibid., Art. 7.

60 Treaty information drawn from UN websites, www.treaties.un.org and www.faolex.org; institutional membership drawn from respective websites, www.fao.org, www.comesa.int, www.eac.int, www.commissionoceanindien.org, and www.iora.int.

61 IORA, 'Overview', , accessed 17 July 2020, www.iora.int/en/priorities-focus-areas/overview.

62 COMESA, 'Fisheries Program', accessed 30 July 2020, www.comesa.int/comesa-fisheries-program/.

63 See www.fao.org/3/a-br824e.pdf.

64 Although the Kenyan National Ocean and Fisheries Policy does refer to 'increased and sustainable fish production and utilization', 3.1.2. See also 3.3.1 where reference is made to prioritizing fisheries resources, conservation and management. The Tanzanian National Five-year Development Plan refers to an abundance of natural resources allowing for development of the fisheries sector, 33.

65 See, for example, Mauritius's National Biodiversity Strategy and Action Plan, which refers to the 'long-term goal of creating a sustainable ocean economy', 1.2 and 2.3.

66 Seychelles' Fisheries Sector Policy and Strategy, 5.

67 Collated from www.faolex.org.

68 See the Kenyan National Ocean Fisheries Policy, paras 3.2 and 3.3; Seychelles' Fisheries Sector Policy and Strategy, 11; and the Tanzanian National Five-Year Development Plan.

69 See, for example, the South African General Policy on the Allocation and Management of Fishing Rights, para. 3.1(a); the Kenyan National Ocean Fisheries Policy, paras 4.2.1 and 5.5.3; Seychelles' Fisheries Sector Policy and Strategy, 11. Interestingly, the Tanzanian National Fisheries Policy refers to the 'rational utilization for sustainable development' of fisheries resources as a policy objective: 13.

70 Seychelles' Fisheries Sector Policy and Strategy refers to explicitly to the CoC-RF: 12. The Somalian National Development Policy refers to the SDGs in detail.

71 The Mauritian National Biodiversity Strategy and Action Plan includes these three aspects of sustainable development in its key principles: para. 2.3, 52. Seychelles' Fisheries Sector Policy and Strategy refers to equity, participation and development of human resources as key objectives alongside economic and sustainability goals: 11. Mauritius has recognized the social and economic importance of marine ecosystems in its Action Plan: para. 1.1.1.2, 22. The Tanzanian National Fisheries Policy refers to 'sustainable … and more efficient commercialized fisheries … that contribute to the improvement of livelihoods and the national economy while conserving environment': 4. The Tanzanian National Five-Year Development Plan links industrialization and human development: 37–39.

72 General Policy on Allocation and Management of Fishing Rights in South Africa (2013) 4.1.5(b).

73 General Policy para. 5.3.5. See also para. 9.1, which commits the nation to ongoing review of the policy to address socio-economic needs of fishing communities and to achieve ecological sustainability of resources. Tanzania includes gender equity in access to fisheries resources and benefits as a policy goal: National Five-Year Development Plan, 31.

74 Seychelles' Fisheries Sector Policy and Strategy defines the fisheries sector as including industrial, small-scale and aquaculture fisheries: 6. The Tanzanian National Five-Year Development Plan includes the empowerment of artisanal fishers (pp. 149, 169 and 252) and the National Fisheries Policy notes the importance of investment in infrastructure for both commercial and artisanal sub-sectors (20).

75 Policy for the Small-Scale Fisheries Sector in South Africa, para. 3.2. See also para. 3.1.

76 See, for example, the Tanzanian Five-Year National Development Plan, 50 and 167, and its National Fisheries Policy, 8.

77 The Tanzanian National Five-Year Development Plan refers to the prioritization of fishing and aquaculture which may be accelerated through interventions such as 'empowering fishers through subsidies': 56.

78 South Africa, for example, makes only passing reference to aquaculture development in the Small-Scale Fisheries Policy, paras 1.4 and 4.1.2; and a single reference is made in Somalia's National Development Plan 5.4.6, 73.

79 The Kenyan Fisheries Sector Policy and Strategy, paras 4.3 and 5.3. See also para. 5.1.4.

80 See pp. 168–69. See also the Tanzanian National Biodiversity Strategy and Action Plan 2015–2020, which refers to aquaculture development and sustainable technologies.

81 National Fisheries Policy, 2–5.

82 Fisheries Sector Policy and Strategy, 23–24.

83 Ibid., 23.

84 For example, the Kenyan Fisheries Service.
85 Including application and decision-making processes, and conditions for approval of licences to fish, registration of fishers and fishing vessels, etc.
86 Such as fishery management plans, conservation measures (including the declaration of aquatic reserves and listing of protected species),
87 Including monitoring and reporting obligations.
88 Offences and penalty provisions.
89 The Kenyan Fisheries Law, Parts VI and XVIII; the Mauritian Fisheries Law, Part V;
90 The Kenyan Fisheries Law Part VII.
91 The Kenyan Fisheries Law, Parts X and XII; the Mauritian Fisheries Law, Parts VI and VII.
92 The Tanzanian Fisheries Law, Part VII.
93 The Kenyan Fisheries Management and Development Act s.5 notes that the 'objective of this Act is to protect, manage, use and develop the aquatic resources in a manner which is consistent with ecologically sustainable development, to uplift the living standards of the fishing communities and to introduce fishing to traditionally non-fishing communities and to enhance food security'.
94 See, for example, the Kenyan Fisheries Management and Development Act s.5 which includes a long list of detailed principles.
95 Fisheries Management and Development Act s.9. See also s.17 on management and control measures.
96 Sea Fishery Act s.2.
97 Marine Living Resources Act s.2.
98 Ibid.
99 Ibid., Parts VIII and X.
100 Ibid., s.11.
101 Ibid., s.11(1)(b).
102 Ibid. s.11(1)(b).
103 Ss.18 and 24.
104 Part III, ss.8–10.
105 Environmental Protection Act 2002 (as amended in 2008) Fifth Schedule.
106 See pp. 187–93.
107 The Tanzanian National Five-Year Development Plan refers to the SDGs, but interestingly not to SDG 14: 41.
108 See, for example, the Mauritian National Biodiversity Strategy and Action Plan para. 3.6.6, 45. This refers to the need to apply the goals of Aichi Target 4 – no net loss of biodiversity – to aquaculture and fisheries operations.
109 Explicit in the South African General Policy on the Allocation and Management of Fishing Rights, for example.
110 See p. 13. See also p. 24 where the linkage between different sectors of the blue economy – fisheries, environment and tourism – is made.
111 In relation to South Africa, for example, see http://archive.iwlearn.net/bclme.org/projects/docs/Final%20report%20LMR-MC-03-01.pdf.
112 Uncoordinated law and over-regulation have been identified as hampering the development of aquaculture in South Africa: Department of Agriculture, Forestry and Fisheries, *Legal Guide for the Aquaculture Sector in South Africa* (2013), 7, www.nda.agric.za/doaDev/sideMenu/fisheries/03_areasofwork/Aquaculture/AquaPolGuidLeg/AquaPolicyGuide/The%20User%20Friendly%20Legal%20Guideline%20for%20the%20Aquaculture%20Sector%20in%20South%20Af….pdf.

4 Coastal and offshore energy and mineral resources

Vivian Louis Forbes, PhD

Introduction

Sub-Saharan Africa is a major exporter of a wide variety of coastal and sea-based minerals and hydrocarbon resources. The region is considered to be the last energy frontier, whose hydrocarbon reserves are expected to turn it into a new global nucleus.[1] The incorporation of hydrocarbon exploration approaches and imaginative ideas, assisted by modern technologies, has yielded significant scientific and economic successes for African countries' blue economies.[2] The focus of attention for oil and gas exploration is on the off-shore sedimentary deposits that remain relatively untapped and the huge wealth of undiscovered hydrocarbon reserves. Africa has witnessed an increase in drilling activities, and in 2019 there were at least 100 oil and gas rigs in place around the continent (compared to 77 in 2017).[3]

The race to get ahead in the deep-water exploration and exploitation of hydrocarbon resources resumed in earnest off western and south-western Africa in the early 2000s and continued to get keener as the decade wore on. Most of these activities took place off Angola and Nigeria where over five billion barrels of crude oil were extracted during the decade. There was sufficient evidence to suggest that the deep waters of the Atlantic and Indian Oceans held prospective reserves even larger than those found in areas that had already yielded relatively large discoveries.[4]

Coastal and offshore oil extraction and processing has been one of the largest areas of foreign direct investment in sub-Saharan Africa roughly since 2005 which gives countries like Kenya, Madagascar and Somalia, which have huge potential to develop their oil and gas industries, some cause for optimism. The Niger Delta is a major giant hydrocarbon province that is classified as an 'Atlantic-type' basin and is one of the world's 78 giant oilfields. With a surface area of 155,400 sq km the Niger Delta oilfield is estimated to incorporate six giant fields with proven reserves of 19,000,000,000 billion barrels.[5] A giant oilfield is deemed to be a reservoir containing 500 million barrels of recoverable oil or gas equivalent. The possession of a giant oil-field – or any mineral resource, for that matter – is potentially promising for a country's economy; however, there are many obstacles to overcome before that potential can be realized and profits accrue to the coffers of the government, operators, international partners and shareholders (sometimes known as the 'resources curse').[6]

Along Africa's eastern continental margin, exploration for hydrocarbon resources was fairly limited until 2010 with fewer than 60 offshore wells drilled before the discoveries in the Mafia Basin and Rovuma Delta located between Tanzania and Mozambique. Indeed, discoveries of large hydrocarbon reserves continue to be made mostly in new and emerging reserves which are usually in the possession of major Chinese, European and US exploration companies.[7]

Since 2017 there has been a burst of exploration activity – both onshore and offshore – within the continental margins of eastern, western and southern sub-Saharan Africa. Significantly, oil seeps and oil indications in wells have been reported within the South African Karoo coalfields; however, very little exploration has taken place in the onshore Karoo Basin. Nevertheless, exploration for, and exploitation of, hydrocarbon reserves are ongoing especially on the strength of promising discoveries of gas along the East African coast and along the north-west continental margins of sub-Saharan Africa.[8]

An optimistic assessment, in a report published in 2020, was the potentially ultra-high impact wells off sub-Saharan Africa.[9] For example, Total Oil's 2019 gas condensate discovery off South Africa prompted the company to explore four more wells in the region during 2020.[10] Off the coast of Namibia, companies including Total Oil, Shell and Kosmos were testing the waters for 'giant fields' during 2020.[11] The same report suggested that 500–600 wildcats were expected to be completed during 2020 worldwide. This development could add around 15 billion barrels of oil equivalent resources, in line with the industry's performance since 2014. Associated investments were expected to reach US $25–$30 billion, a similar amount to that spent in 2019, although the figure could be 5%–15% lower if cost efficiencies continue.[12]

In 2015 Africa's proven oil and gas reserves accounted for 7.5% and 7.1% of global reserves, respectively (over 1,698 thousand million barrels of oil and about 187 trillion cu m of gas). As international energy prices recovered, the oil producing nations attracted investor interest. Oil and gas discoveries during 2018–19 coupled with regulatory changes and fast-growing energy demand from expanding local consumer markets offered significant opportunities across the continent. In a 2018 report entitled *African Oil and Gas State of Play*, Deloitte examined notable destinations in nine sub-Saharan African countries, eight of which are located on the coast: Angola, Congo-Brazzaville, Equatorial Guinea, Gabon, Ghana, Mozambique, Nigeria, Senegal and Tanzania. Nigeria is the region's top producer, has the largest proven oil reserves and will contribute 46% of the region's crude production by 2025, according to Global Data.[13]

Hydrocarbon province

Since the early 1990s, there has been a proliferation of exploration for hydrocarbon reserves in 'frontier provinces' (the term 'province' was deemed preferable to 'basins' because many of the areas are not basins in the truest sense). Hydrocarbon reserves have been discovered in overthrust or fold belts, cratonic arches and fan deposits. Deep-sea fans (such as the Congo Fan, the Niger Delta Fan and the Zambesi Delta Fan) and cones are included as are carbonate bank build-ups on elevated oceanic crusts. These latter provinces were rated as having 'poor' potential; however, the technology for recovering hydrocarbon resources continues to yield surprises.

New data, greater knowledge and understanding, and better and accurate interpretations of drill log and satellite-derived and computer-enhanced images has benefited the oil industry in its quest to discover new reserves. The mature petroleum provinces in sub-Saharan Africa lie off the shore of West Africa and the Niger Delta. There are six major sedimentary provinces that front the west coast of Africa, from the Ivory Coast to Angola in the south-west.

In 1999 the East African Rift Hydrogeology Systems (EARHS) Project analysed the hydrocarbon potential of the East African continental margin which covers an area measuring more than 10 million sq km and includes eight countries where there are prospective Phanerozoic sedimentary basins. The major sedimentary provinces along the eastern seaboard of the East African continental margin, from north to south are the Somalia Coastal Basin, the Lamu Basin, the Tanzania Coastal Basin, the Ruvuma Basin and the Mozambique Coastal Basin. On the western margin of the island of Madagascar are the Majunga Basin and the Morondava Basin.[14]

Many of the basins have accumulated sediments to a thickness of 10 km or more.[15] The margin from the Lamu Basin off Kenya's coast in the north to the Zambesi Delta in the south is covered by thick Tertiary and Cretaceous sediment derived from the East African Rift shoulders and Lower Jurassic source rocks are predicted to be in the gas window along most of the margin.[16] The East African margin is a complex structure caused by multiple phases of rifting in different directions. The recent gas discoveries in Mozambique and Tanzania are believed to be sourced from over-mature Jurassic or possibly deeper Permian-era Karoo shales. The margins in South Africa are less deeply buried and have better oil-bearing potential.

Exploration

A 2010 World Bank study focused on the petroleum markets and assessed 12 countries, Burkina Faso, Côte d'Ivoire, Mali, Niger, Senegal in West Africa and Botswana, Kenya, Madagascar, Malawi, South Africa, Tanzania and Uganda in East and Southern Africa.[17] The markets studied were considered 'small', with all but three having domestic consumption of less than 25,000 barrels per day (b/d). Against a world-scale refinery size of at least 100,000 b/d, only three out of seven petroleum refineries in the countries studied were deemed 'world-scale refineries' in size and complexity, and all were located in South Africa. The study suggested that the monitoring and enforcement of rules already in place must be strengthened in most of the countries. Commercial malpractice, in Angola and Nigeria for example, if left unchecked, could take over the market and drive out efficient operators known to refrain from engaging in fraud.[18] America's ExxonMobil is one of the largest foreign investors in Africa and operates in at least 21 countries. It has committed more than US $24 billion to energy exploration and development. US companies have significant investments in Angola, particularly in the energy sector. Tullow Oil operates in four coastal countries. In 2019 a Chevron-led consortium announced plans to invest more than $2 billion to explore new offshore natural gasfields and to increase gas production in existing fields, thanks largely to the Angolan government's economic reforms and efforts to attract investment. The interest of major US energy companies in Africa has not decreased, and, despite a downturn in the market during 2020, the needs of Asia and Europe will not stop growing.[19]

Italy remains a close second as a source of foreign oil activity, operating in at least 12 African countries. ENI, an Italian multinational oil and gas company, planned investments in 2019 representing 60% (around US $25 billion) of the company's total investment mainly in oil and gas. By comparison British Petroleum (UK) planned investments in 2020 amounting to over US$ 2.5 billion.[20]

The People's Republic of China is the world's largest consumer and importer of crude oil, and in 2019 it imported about 10 million b/d.[21] It is estimated that the country imports over 66% of its total oil requirements. In a bid to position themselves as global players in the international oil and gas market, Chinese national oil companies (NOCs) have significant investment plans for 2019–2030 in Africa's oil and gas industry. Africa is the second largest region to supply oil and gas to China, after the Middle East, and exports more than 25% of its total oil and gas to China. China's state-owned enterprises operate in at least eight African nations, namely Angola, Chad, Gabon, Mauritania, Mali, Mozambique, Niger and Nigeria. Chinese NOCs are operating extensively throughout Africa.[22]

A wealth of data currently held with various agencies and available to interpreters as models for exploration has considerably changed our understanding of the geology, evolution, structure and depositional patterns of the continental margin and related sedimentary basins at all scales. The interpretation of satellite-derived gravity data, for example, is enabling the development of basin- or margin-scale models that permit the characterization of the general structure and development of the margin. Additionally, the interpretation of high-resolution 2-D and 3-D seismic data provides a detailed view of the stratigraphy, structures and volcanic bodies comprising these basins.[23]

The sheer wealth of drilling-related and outcrop data for the onshore region takes into account the characterization of the timing and amplitude of tectonic changes along the West Africa margin. The application of remote-sensing technologies and 3-D seismic data allows geoscientists to further interrogate the structural definition of the East and West Africa continental margins. With the rapid development of sensor technology, computing power and its data support, hydrocarbon exploration is now thoroughly digitalized and mature. A smart approach to 'condition-based maintenance' (CBM) is achieved as a result of digitalization, enabling improved lifecycle performance for operations and maintenance.[24]

Production

Africa has an abundance of proven petroleum or crude oil reserves estimated at 125 billion barrels in 2019. They are largely concentrated in the north and also in Western Africa, although further hydrocarbon reserves (oil and gas) are still being identified in new locations in sub-Saharan Africa, and in some cases have already been developed commercially. The regional shares of crude oil production are as follows: Western Africa has 30%, Central Africa about 31%, Eastern Africa about 5% and Southern Africa a mere 1%.[25]

Refining

The trend in the global context since the 1980s has been to replace numerous small simple refineries with fewer, larger and more complex ones. Refinery closure is often politically sensitive, especially if the refinery is government-owned. A burden on the state's economy is the protection of refineries that cannot compete with direct product imports and the effective and sustainable management of its continental margin. Allegedly, millions of dollars of tax-payers' money are spent on oil and gas refineries within sub-Saharan Africa, in the guise of repairs and turn-around maintenance. Yet these refineries are unable to work close to installed capacities, compelling the region to continue importing almost all the refined oil products it consumes.

Table 4.1 Top five oil producers from coastal states in sub-Saharan Africa

Country	*Oil production b/d, 2019*	*World ranking*
1 Nigeria	1,999,885	13
2 Angola	1,769,615	14
3 Congo (Republic of)	308,363	31
4 Equatorial Guinea	227,000	37
5 Gabon	210,820	38

Source: US Energy Information Agency (2019).[26]

Nigeria and South Africa each have six refineries; Angola and Côte d'Ivoire each have two refineries. In each of the 12 sub-Saharan African countries under study in this chapter there are collectively 12 refineries. Dangote Refinery in Nigeria is the largest with a capacity of 650,000 b/d of oil and 2,500 cu m/d of gas. Five of the six refineries in South Africa produce just over 100,000 b/d of oil and an average of 20,000 cu m/d of gas. Refining capacity refers to the given capacity of total crude charge input which a refinery is built to handle before the crude is converted into other consumable products.

In 2019 the global refining industry began to witness a change in investment patterns, with strong growth in investment for capacity building, and for the upgrading and modernization of facilities to produce greater output from lesser inputs, in addition to meeting strict environmental requirements. However, by April some refineries in sub-Saharan Africa had halted production or were considering reducing their throughput due to reduced demand owing to the outbreak of the coronavirus (COVID-19) pandemic and soaring inventories. For example, in South Africa, both Engen and Sasol either shut down their oil refineries temporarily or suspended processing activities. The refinery in Pointe Noire, Republic of the Congo, operated at over 80% of its capacity. Refineries in Angola, the Democratic Republic of the Congo (two), Nigeria (three) and Senegal were undergoing upgrades, and many others were planning to carry out ongoing maintenance during the latter half of 2020. The Nigerian National Petroleum Corporation planned to fully rehabilitate all four of its refineries by January 2021 so that they would be competitive by 2022.[27]

Natural gas

Natural gas consists primarily of methane, a very clean and safe fossil fuel, although the by-products of its combustion, such as carbon dioxide, carbon monoxide and nitrogen oxides, are environmental pollutants and contribute to climate change. On the other hand, these emissions are thought to be almost half that produced by coal use. It is also more economical than coal, as the costs and time periods required to build a gas plant are considerably less than for coal plants.

Nigeria, Cameroon and Algeria also have significant associated natural gas, which has historically been flared in the earlier years of oil production due to the lack of commercial opportunities. This situation is now changing as the value of natural gas has increased since then. Africa is estimated to have 52 trillion cu m (tcm) of remaining recoverable conventional natural gas resources, of which 31 tcm are in sub-Saharan Africa. Proven gas reserves in sub-Saharan Africa have increased by 80% since 2000, of which around 70% is in deep water. One-sixth of sub-Saharan African's proven natural gas reserves are associated with oil. Important natural gas resources have been recently discovered in Mozambique and Tanzania.[28]

Renewable energy resources

Renewable energy, i.e. theoretically inexhaustible energy sources that are not derived from fossil or nuclear fuel, includes biofuels, biomass, geothermal, hydrogen, hydropower, solar, hydrokinetic energy (tides and waves) and wind. These resources are the centre of the transition to a less carbon-intensive and more sustainable energy system, The cost of developing renewable energy technologies remains high, but the increased demand can lead to economies of scale and wider use, especially in developing regions like sub-Sahara Africa where energy demand is rising and many renewable resources are abundant.

In 2018 renewable electricity generation in Africa was 163 Terrawatt hours (TWh) and is expected to rise to 716 TWh by 2030. In the sub-Saharan Africa context the following brief commentary and the tabulation in Annex II may be useful indicators. The share of renewable energy in total final energy consumption in many of the sub-Sahara African countries grew, but in some cases it fell. Most of this energy was provided by solid biofuels and traditional biomass.[29]

Bioenergy is derived from biomass. Fuelwood or wood fuel is the most important energy source across sub-Saharan Africa. Biofuel is combustible matter derived from biomass (such as crops) to generate electricity and power and to produce liquid fuels such as bioethanol and biodiesel.

It has long been recognized that there is significant potential for the further development of hydropower in the region. Some of the main hydroelectric hubs include the Niger and Senegal Rivers and Guinea; the Congo River and Inga scheme; the Nile River development; and in Southern Africa, involving the Orange, Limpopo and Zambesi Rivers.

The marine environment stores hydrokinetic energy in the form of heat, currents, tides and waves which can be utilized as renewable energy. There is an enormous potential for generating power from wave energy, for example, along the coast of Southern Africa. Tidal energy is created through the use of generators. Large underwater turbines are placed in areas with high tidal movements to capture the kinetic motion of the ebbing and surging of ocean tides to produce electricity.

Geothermal energy is the heat contained within the earth that can be recovered in the form of steam and hot water to be utilized for power generation and other direct use applications. Geothermal energy potential that can be used for power generation is mostly concentrated in the East Africa Rift System (EARS). The EARS cuts through Eritrea, Djibouti, Ethiopia, Kenya, Tanzania, Uganda, Rwanda, the DRC, Zambia, Malawi, Mozambique and Madagascar. Further geothermal developments are underway in Djibouti, Eritrea, Ethiopia, Kenya, Tanzania and Uganda, which aim to increase the generating capacity of these countries.

Solar energy is produced through a process called photovoltaic effect, whereby solid-state electronic cells are used to produce a direct electrical

Table 4.2 Renewable energy resources

Country	*Biomass*	*Hydro*	*Geothermal*	*Tidal/wave*	*Solar*	*Wind*	*Renewable energy consumption (% of total energy consumption)*
Angola	extensive	excellent				feasible	57%
Benin	extensive	potential	low		good	unknown	51%
Cameroon	3rd, SSA	4th, Africa	no effort		good	feasible	78%
Cabo Verde	low	low	limited		high	excellent	21%
Comoros	extensive	modest	high		good	low	47%
Côte d'Ivoire	extensive	modest	limited			modest	74%
Congo, Democratic Republic of	high	good	potential		high	low	96%
Congo, Republic of	good	good	unknown		unknown	good	48%
Djibouti	limited	nil	potential		high	feasible	34%
Equatorial Guinea	good	low	unknown		low	nil	30%
Eritrea	extensive	low	potential		moderate	potential	80%
Ethiopia	extensive	potential	potential		potential	good	94%
Gabon	good	good	unknown		moderate	potential	70%
Gambia, The	extensive	nil	unknown		good	good	49%
Ghana	extensive	good	unknown		good	good	49%
Guinea	extensive	potential	unknown	potential	low	low	76%

Country	*Biomass*	*Hydro*	*Geothermal*	*Tidal/wave*	*Solar*	*Wind*	*Renewable energy consumption (% of total energy consumption)*
Guinea- Bissau	extensive	potential	viable		minimal	good	89%
Kenya	moderate	unstable	good		good	good	78
Liberia	extensive	potential	unknown		moderate	good	89%
Madagascar	extensive	modest	potential			good	79%
Mauritius	extensive	modest			good	good	modest
Mozambique	extensive	potential			unknown	unknown	88%
Namibia	moderate	vibrant	unknown		good	excellent	33%
Nigeria	extensive	unknown	good		potential	good	86%
São Tomé and Príncipe	extensive	potential	potential		potential	inadequate	43%
Senegal	extensive	good	unknown		potential	adequate	51%
Seychelles	potential	nil	nil		potential	good	modest
Sierra Leone	extensive	good	nil		potential	unknown	80%
Somalia	extensive	low	poor		potential	good	94%
South Africa	minimal	potential	modest	potential	good	excellent	modest
Tanzania	excellent	good	potential		good	good	88%

Source: Extracted by the author from various publications (see References in this chapter).

Note: The description offered is qualitative.

current from the sun's radiant energy. Solar cells are used to generate electricity for household power, street lighting, highway telephones, calculators, watches and water pumping, among other applications.

Wind energy is one of the world's fastest growing sources of renewable energy. The majority of windmills have a horizontal axis, but some have a vertical axis. A single wind machine can produce between 1.5 million kWh and 4.0 million kWh of electricity per year. Wind speed increases with altitude and over open areas with no windbreaks.[30]

Offshore and deep-sea mining

The existence of huge offshore mining reserves and the prospect of finding new hydrocarbon deposits prompted over 70 member states to make submissions to the UN Convention on the Law of the Sea's Commission on the Limits of the Continental Shelf, seeking recognition of their claims over areas of the continental shelf. The applications involved claims to areas between the boundaries of their exclusive economic zones (usually 200 nautical miles or 322 km from their territorial sea baseline system) and a further boundary of a maximum 350 miles (563 km), depending on the length of the continental margin.[31]

These deep-sea reserves include minerals that can be found in the form of polymetallic nodules or polymetallic sulphides, which are contained in metalliferous muds. Cobalt-rich ferromanganese crusts are found at shallower depths of between 400 m and 5,000 m in areas of volcanic activity, with some located in the Comoros archipelago or around Madagascar. These crusts also contain iron oxide, molybdenum, zirconium or rare earths. To date, the International Seabed Authority (ISA) has issued only 27 exploration permits around the world.

Despite environmental concerns, deep-sea mining will be a reality by 2030 and African coastal and island states need to ensure that they will benefit from the fortunes that can be found on the seabed. In February 2018, at the ISA meeting, the African Group called for a voluntary scheme to underwrite developing countries participating in the organization. Many environmental issues need to be resolved before extensive seabed mining takes place. The China Ocean Mineral Research Development Association, a Chinese state agency, was granted a ferromanganese block near the Madagascar plateau.[32]

Deep beneath the frigid waters of the Atlantic Ocean off the coast of Namibia, some of the world's most valuable diamonds are strewn across or buried just beneath the seabed. These 'maritime mines' may extend a lifeline to countries like Namibia whose economies depend on diamonds. In 2019 mining companies allegedly extracted US $600 million worth of diamonds off the Namibian coast, sucking them up in giant vacuum-like hoses. In 1991 De Beers, which has dominated global diamond production, purchased mining rights to more than 3,000 sq nautical miles of the Namibian seafloor. By 2019 the mining company had explored only 3% of that area. More countries are encouraging exploration to commence along their coastlines. Sailing under the flag of Namibia, the SS *Nujoma* is the world's most advanced deep-water diamond exploration vessel, and sucks up sediment from the ocean floor to find diamonds. More than 90% of Namibia's diamond-related revenue comes from offshore finds. Most of the diamonds are apparently located close to the surface of the seabed.[33]

It is beyond the scope of this chapter to provide a more comprehensive coverage of the precious mineral wealth of Sub-Saharan African countries. However, it must be stated that an abundance of valuable coastal minerals, including molybdenum and vanadium, have been discovered along the coast of south-west Africa in several shallow basins. Furthermore, vast quantities of ocean phosphorite have been located on the continental margin (and on the abyssal ocean floor).[34]

Challenges

The fundamental challenge, of course, is the intrinsic volatility of the energy sector. Producers need time to address the vagaries of an over- or under-supplied market. They also need to grapple with the pace and magnitude of the transition to generating energy from non-fossil fuel sources. In the face of these uncertainties, oil and gas companies must develop a resilient strategy to mitigate these risks. Another challenge confronting the industry is supply disruption. In existing oilfields, production is declining – and this decline rate is accelerating by about 4% per annum. Current spending increases elsewhere are insufficient to ensure discovery of enough new fields to replenish this decline.[35]

Nuclear power is generated through the use of uranium or plutonium. The production of uranium generates a huge amount of radioactive waste in the process. Uranium is enriched in nuclear reactors that generate heat through nuclear fission and steam to drive turbines and generators. Most nuclear plants are situated along the coast so that seawater can be used as a cooling mechanism instead of cooling towers.

South Africa is the only nation in Africa currently producing electricity from nuclear sources, with nearly 14,202 GWh produced in 2013. A number of other countries, including Kenya and Namibia, have expressed an interest in introducing nuclear power into their domestic energy mix. Nigeria has developed and put in place the required institutional and regulatory framework for the successful implementation of nuclear power programme as required by the International Atomic Energy Agency. Three Sub-Saharan African countries, Namibia, Niger and South Africa, are among the world's 10 biggest holders of uranium reserves. Africa provides a significant share of global production (18%) of uranium: Namibia (8.2%), Niger 7.7%, Malawi (1.2%) and South Africa (1.1%).[36]

Generally, foreign oil companies prefer to enter into negotiations relating to exploration and exploitation of hydrocarbon resources with governments that have a petroleum code catering for several types of petroleum contracts that offer incentives. These include agreements attached to exploration permits or exploitation concessions; production-sharing contracts; other agreements such as risk services contracts; and additional investment credits offered for exploration in deep and ultra-deep waters.[37] For example, Nigeria's key energy legislation in this regard is the Petroleum Industry Governance Bill. The legislation is intended to create efficient governing institutions with clear objectives; however, its eight-year passage through parliament has been hindered by internal political wrangling and objections from foreign oil companies who are unhappy about the significantly higher taxes that were proposed in earlier drafts.

During the Nigerian International Petroleum Summit held in Abuja in 2020, the Nigerian government sought to reassure delegates about the country's openness to investments and efforts to bolster the regulatory framework. These include attempting to pass the Petroleum Industry Bill into law during 2020.[38]

Few governments in the region make key sector data regularly available in a timely manner. Such information encourages consumers and enables informed debates about prices and sector efficiency. Making price and other data widely available has taken on greater importance since 2017 against the backdrop of volatile oil prices and calls from different sectors in many countries for greater price control to protect consumers.

International oil companies are generally wary of committing funds in expensive deep-water frontier exploration in countries that have experienced political instability especially those with long periods, for example, as in the case of Côte d'Ivoire and its armed rebellion of 2002 to the end of the civil war in April 2011. The instability of the government no doubt resulted in a slow-down in investment in the oil and gas sector.[39]

Conclusion

Coastal and maritime energy resources – the foundations of the blue economy – are fundamental to national development and are vital to economic growth and progress. Energy resources are more than a matter of technical recovery. International oil exploring and producing companies have developed a partnership-based model, working with countries around the world. Offshore final investment decisions are not easily made: the decision to drill generally takes many years. Energy production companies work with governments to understand their self-determined energy ambitions and self-interest.

Poverty and cronyism persisted in most countries which were too reliant on oil because of the 'resources curse'. However, many countries in the region realize the importance of appropriate legislation and the need to enforce rules and regulations. Thus, Sub-Saharan African countries will need to revisit their existing national legislation and energy and marine resources policies if they have not already done so.

Sub-Saharan African countries are committed to various energy-related goals and actions to achieve the post-2015 Agenda at global, continental and national levels. These include the African Union's Agenda 2063, which commits to harnessing Africa's energy resources to ensure modern, efficient, reliable, cost-effective, renewable and environmentally friendly energy supplies for all people of Africa; the African Development Bank's very ambitious goal to achieve universal access by 2025; and, the Sustainable Development Goals (SDGs), which Agenda 2063 echoes. In the context of the SDGs, parties have committed to ensuring access to affordable, reliable, sustainable and modern energy for all while urging countries to combat climate change and its impacts.

Notes

1 United Nations Environment Programme (2017) 'Atlas of Africa Energy Resources', African Development Bank 2017, www.unenvironment.org/resources/report/atlas-africa-energy-resources; US EIA (2020) 'Short-Term Energy Outlook', June, Washington, DC: US Energy Information Administration, www.eia.gov/outlooks/steo/outlook.php; ibid. (2020) 'International Energy Outlook 2019, 24 September, www.eia.gov/outlooks/steo/outlook.php; Africa-EU Energy Partnership (AEEP) (2016) *Africa-EU Energy Partnership Status Report Update 2016*, EUEI: Eschborn; AfDB (2014) 'Africa Energy Sector: Outlook 2040', www.euei-pdf.org/en/status-report-africa-eu-energy-partnership; African Development Bank (AfDB) (2016) *Programme for Infrastructure Development in Africa* (PIDA), Abidjan: AfDB, www.euei-pdf.org/en/status-report-africa-eu-energy-partnership; ibid., *The Bank Group Strategy for the New Deal on Energy for Africa 2016-2025*, Abidjan: AfDB, www.afdb.org/fileadmin/uploads/afdb/Documents/Generic-Documents/Bank_s_strategy_for_New_Energy_on_Energy_for_Africa_EN.pdf; African Energy Commission (AFREC) *Africa Energy Database Edition 2019*, https://au-afrec.org/publications/afrec-africa-energy-database-2019-en-fr.pdfritish; British Petroleum (BP) (2016) *BP Statistical Review of World Energy*, June, London: BP, www.bp.com/en/global/corporate/energy-economics/statistical-review-of-world-energy.html; Desertec-Africa, Africa Is Endowed with Huge Energy Resources, Desertec-Africa, www.desertecafrica.org/ (accessed 29 January 2016); DLIST (Distance Learning and Information Sharing Tool) Benguela, 'Energy Sources: What Are the Pros and Cons', United Nations Development Programme (UNDP), http://archive.iwlearn.net/dlist-benguela.org/Burning_Issues/Energy/Energy; International Energy Agency (IEA) (2015) *World Energy Outlook 2015 (WEO-2015)*, Paris: OECD/IEA, www.iea.org/reports/world-energy-outlook-2015; KPMG Africa (2015) *2015 Sector Report on the Oil and Gas Sector in Africa*, KPMG International Cooperative; https://assets.kpmg/content/dam/kpmg/pdf/2015/07/ru-en-oil-and-gas.pdf.

2 US Energy Information Administration (EIA) especially studies relating to energy analyses of African countries, www.eia.gov/; Richard Barrett, Martijn Van Haaster, Susan Witte, Stephen Collins, Stephano Baffi and Shell E&P International Rijswijk, Netherlands, 'The Evolution of East Africa Passive Margin and Petroleum Systems', AAPG International Conference, Barcelona, 21–24 September 2003; Morgan Brazilian, Patrick Nussbaumer, Hans-Holger Rogner, Abeeku Brew-Hammond, Vivien Foster, Shonali Pachauri, Eric Williams, Mark Howells, Philippe Niyongabo, Lawrence Musaba, Brian O'Gallachoir, Mark Radka and

Daniel M Kammen (2012) 'Energy Access Scenarios to 2030 for the Power Sector in sub-Sharan Africa', *Utilities Policy*, 20, pp. 1–16.

3 'Africa's Oil and Gas Scene after the Boom: What Lies Ahead', *Forum*, no. 117, Jan. 2019, www.oxfordenergy.org/wpcms/wp-content/uploads/2019/01/OEF-117.pdf; World Energy Council (2016) *World Energy Resources 2015*, London: World Energy Council, www.worldenergy.org/publications/entry/world-energy-resources-2016.

4 See Global Exploration and Production, 1985–2020E; www.in-vr.co/global-ep; BHP (2017) 'The Future of Global Exploration and Production'.

5 Jim Brooks (1988) 'Classic Petroleum Provinces', *Geological Society Special Publication*, No. 50, pp. 1–8.

6 S. W. Carmalt and B. St John (1986) 'Giant Oil and Gas Fields', in M. T. Halbouty (ed.) *Future Petroleum Provinces of the World*, Tulsa: AAPG, pp. 11–53.

7 R. Nehring (1978) *Giant Oil Fields and World Oil Resources*, Rand Corporation Report R-2284 CIA, www.rand.org/pubs/reports/R2284.html; B. St John (1980) *Sedimentary Basins of the World and Giant Hydrocarbon Accumulations*, American Association of Petroleum Geologists, Special Publication; Robert E. Mattick, Frank D. Spencer and Frederick N. Zihlman (1982) 'Assessment of the Petroleum, Coal and Geothermal Resources of the Economic Community of West African States (ECOWAS) Region', doi:10.2172/5585730.

8 *Oil and Gas Journal*, various issues in 2019–21 offer commentary on promising discoveries and related issues such as drilling and production, refining and processing, pipelines and transportation, COVID-19, LNG and White Papers on a regular basis. For example, 'IEA: Oil Fundamentals on Strong Trajectory' 19 January 2021, www.ogj.com/general-interest/economics-markets/article/14195815/iea-oil-fundamenhttps.

9 Wood McKenzie (2020) 'Global Oil Supply Short-term Update', 9 January, and 'Oil Product Price Forecast Mid-Month Update January 2020', 20 January, www.woodmac.com/reports/refining-and-oil-products-oil-products-price-forecast-mid-mon.

10 *Offshore Online* (2020) 'Total Oil Discovery off South Africa', 22 January, www.offshore-mag.com/subsea/article/14196158/ten.

11 *Offshore Online* (2020) 'Total Oil, Shell and Kosmos to Test Giant Fields', 27 February, www.offshore-mag.com/subsea/article.

12 Wood McKenzie, Global Oil Supply', and 'Oil Product Price Forecast'.

13 Deloitte (2018) African Oil and Gas State of Play, November 2018; UNEP, 'Atlas of Africa Energy Resources', pp. 2, 3.

14 L. C. Nguyen *et al.*, 'Reconstruction of the East Africa and Antarctica Continental Margin', *Journal of Geophysical Research: Solid Earth*, 4156–79; M. P. Hochstein (1999) 'Geothermal Systems along the West African Rift', *Bulletin d'Hydrogéologie*, No. 17.

15 F. G. Rayer and O. L. Slind (1999) 'The EARHS Project', *Offshore*, 1 February, www.offshore-mag.com/; 'Hydrocarbon East Africa', *Oil and Gas Journal*, 1999, www.ogj.com/.

16 E. Seibold and D. Futter (1982) 'Sediment Dynamics on the Northwest African Continental Margin', in E. Scrutton and M. Tulwani, *The Ocean Floor*, J. Wiley and Sons, pp. 147–63; Ian Davidson and Ian Steel (2018) 'Geology and Hydrocarbon Potential of the East African Continental Margin: A Review', The Geological Society, *Petroleum Geoscience*, 34 (1), pp. 57–91, www.earthdoc.org/content/journals/10.1144/petgeo2017-028.

17 Masami Kojima, William Matthews and Fred Sexsmith (2010) *Petroleum Markets in Sub-Saharan Africa: Analysis and Assessment of 12 Countries*, Extractive Industries and Development series, no. 15, Washington, DC. World Bank; Davidson and Steel, 'Geology and Hydrocarbon Potential'.

18 World Bank (2010); World Bank (2018) *Africa's Pulse*, vol. 17, April, https://openknowledge.worldbank.org/handle/10986/29667.

19 US Department of State, Daily Reports online, 7 March 2020, https://home.treasury.gov/about/offices/international-affairs.

20 Government of the UK, Ministry of Finance (2020) UK-Africa Trade and Investment, 23 January, www.gov.uk/government/topical-events/uk-africa-investment-summit-2020; BP (UK) 'Upstream Major Projects Investments', www.bp.com/en/global/corporate/investors/upstream-major-projects.html.

21 EIA (2020) *Annual Energy Outlook* 2020 with Projections to 2050, 20 March, Washington, DC: EIA, www.eia.gov/outlooks/steo/outlook.php.

22 Institute of Developing Economies-Japan External Trade Organ (IDE-JETRO), www.jetro.go.jp/en/jetro/activities/ide.html (accessed 26 June 2020); Michelle Benavidez, 'Sub-Saharan African Oil: Connections with China', *Pangea Journal*, https://sites.stedwards.edu/pangaea/volume-5-table-of-contents/sub-saharan-african-oil-connections-with-china/.

23 Teresa S. Ceraldi, Richard Hodgkinson and G. Backe (2016) *The Geology of the West Africa Continental Margin: An Introduction*, London: Geological Society, August, doi:10.1144/SP438.11, www.researchgate.net/publication/306089596_The_petroleum_geology_of_the_West_Africa_margin_An_introduction (accessed 17 November 2019).

24 'A Complete Guide to Condition-Based Maintenance' (CBM) in the Context of Computer Software in Industry, https://limblecmms.com/blog/condition-based-maintenance/.

25 UNEP, *Atlas of Africa Energy Resources*; E. Graham and J. S. Ovadia, 'Oil Exploration and Production in S-S Africa, 1990-Present', *The Extractive Industries and Society*, https://researchgate.net/publication (accessed 20 June 2020).

26 US EIA, 'International Energy Outlook 2019'.

27 *Oil and Gas Journal* (2019; 'Refineries', www.ogj.com/general-interest/economics-markets/article/14195815/iea-oil-fundamenhttps.

28 [28] OECD/IEA (2014) *Africa Energy Outlook: A Focus on Energy Prospects in Sub-Saharan Africa*. Paris: OECD/IEA, www.iea.org/reports/world-energy-outlook-2014; World Bank (2016) Renewable Electricity Output (% of Total Electricity Output), http://data.worldbank.org/indicator/EG.ELC.RNEW.ZS.

29 UNEP, *Atlas of Africa Energy Resources*; Alexandros Gasparatos, Lisa Y. Lee, Graham P. von Maltitz, Manu V. Mathai, J. A. P. de Oliveira and Katherine J. Willis, 'Biofuels in Africa: Impacts on Ecosystem Services, Biodiversity and Human Well-Being', Institute of Advanced Studies (IAS), Oxford: United Nations University (UNU), http://collections.unu.edu/eserv/UNU:2902/Biofuels_in_Africa1.pdf; International Energy Agency (2018) 'Renewable Energy Policy Network for 21st Century, (REN 21) IEA/IRENA, www.iea.org/policies/5101-renewable-energy-policy-network-for-the-21st-century-ren21.

30 DLIST Benguela; M. A. Maehlum (2013) 'Wave Energy Pros and Cons', 3 May, http://energyinformative.org/wave-energy-pros-and-cons/ (accessed 11 March 2016); A. A. Mshandete and W. Parawira (2009) 'Biomass Technology Research in Select sub-Saharan African Countries: A Review', *African Journal of Biotechnology*, 8(2), pp. 116–259.

31 UN Law of the Sea Convention 1982. Provisions for maritime jurisdiction limits contained in Articles, 2, 3, 32, 55, 57, 76, 83. Full text available in six languages, www.un.org/Depts/los/convention_agreements/texts/unclos/unclos_e.pdf.

32 International Seabed Authority (2020) *Study of Potential Impact of Polymetallic Nodules*; China Ocean Mineral Resources (2017) http://isa.org.im/map/china-ocean.

33 Instability of governments and international oil companies are of concern in many regions of the world, for example, in the Persian Gulf, the Timor Sea and South China Sea. See, for example, V. L. Forbes (2020) 'Maritime Boundary Delimitations in the Timor Sea: Australia's Experience during Five Decades', No. 96, CDiSS Commentary, 23 December, https://cdiss.commentary.upnm.edu.my

and V. L. Forbes and B. A. Hamzah (2021) 'China's Luconia Shoals Gambit Part of a Larger Picture', *Asia Sentinel Online*, 1 February, www.asiasentinel.com/p/chinas-Luconia-shoals.

34 See Chapter 2, 'Phosphorite on the Ocean Shelves', in *Developments in Sedimentology*, ed. G. N. Baturin, vol. 33, 1981, Elsevier, pp. 55–161.

35 Nehring 'Giant Oil Fields and World Oil Resources'; St John '*Sedimentary Basins of the World*; Mattick et al. 'Assessment of the Petroleum, Coal and Geothermal Resources'.

36 Nigeria's Petroleum Code; See the *Oil and Gas Law Review*, edn 7, October 2019, https://thelawreviews.co.uk; AAPEA 'New Oil and Gas Investment Needs Policy Stability', 8 May 2020; Kristen Bindermann (1999) *Production Sharing Agreements: An Economic Analysis*, Oxford: Oxford Institute for Energy Studies, see e.g. Angola, pp. 70–76.

37 The Nigerian International Petroleum Summit, Abuja, 2020, *Oil and Gas Journal* .

38 'Ivory Coast Aims to Double Oil and Gas Output by 2020', http://reuters.com/article…

39 IEA, *World Energy Outlook 2015*.

5 Maritime safety and security

Jade Lindley, PhD

Introduction

The maritime region surrounding sub-Saharan Africa is vast: it links many important trade routes to major partners in Europe, North America and Asia; it is an underwater flora and fauna mecca for tourism opportunity; it provides protein supporting food security for the region and beyond; and it is resource and energy rich. Ensuring security over the region's blue economy is therefore critical. Maritime security is the absence of criminal threats. While definitions vary, experts agree that criminal threats undermine maritime security and may include maritime terrorism, piracy, trafficking of narcotics, people and illicit goods and weapons, illegal fishing and environmental crimes (Beuger, 2015; Bradford, 2005; Klein *et al.*, 2010; Roach, 2004; Secretary-General of the United Nations, 2008). Maritime *in*security prevents achievement of blue economy goals.

The extensive sub-Saharan African coastline can be protected within each state's declared 200-nautical-mile exclusive economic zone (EEZ) as defined by the United Nations Convention on the Law of the Sea (UNCLOS) (United Nations, 1982: Article 48). Beyond the EEZ exists expansive international waters that may facilitate offending due to natural isolation and, thus, lack of surveillance. These crimes may threaten blue economy goals and therefore there is a strong argument for threats within it, both home-grown and foreign, to be dealt with at the regional level.

Furthermore, high poverty and under- or unemployment may lead many into unscrupulous activity in coastal and maritime areas. Central to the blue economy agenda is the creation of meaningful employment and poverty reduction (ECOSOC, 2018). Maritime insecurity may start on land and weakened governance at the domestic level may enable it to spill over into the maritime domain. This can be seen in the increase in maritime piracy off the coasts of Somalia between 2008 and 2013 and Nigeria since 2011 (Brume-Eruagbere, 2017; Lindley, 2016). While maintaining and respecting governance of each state's declared EEZ, it is important to respond regionally to protect the collective blue economy from criminal threats that undermine regional and even global maritime security. Underpinned by United

Nations (UN) Sustainable Development Goal (SDG) 16, there is a regional responsibility to respond to crimes that hinder blue economy goals, such as organized crime (16.4) and corruption (16.5), and seek to improve the rule of law (16.3) (UN, 2015: 25).

Given the link between land and sea, understanding what causes domestic weakening is part of the puzzle. For example, weak governance in developing, post-conflict, fragile and failing states creates vulnerabilities on land that may therefore permeate into criminal threats at sea. Threats may present in several ways, including for example as economic and social threats causing increased vulnerability among citizens exposed to these security challenges, leading to, or enabling crime. Several sub-Saharan African littoral states such as Yemen, Somalia, Sudan, Nigeria and Cameroon are developing or fragile (Fragile State Index, 2019), and therefore weak enforcement of maritime crimes in and around the EEZ may be problematic at the domestic and regional level. The vast oceanscape provides opportunities to engage in various crimes (UN, 2004). This is particularly important given that maritime criminals often operate across overlapping territorial and international waters, linking to 'ports of convenience' whereby corruption enables criminal activities (Petrossian, 2014). Livelihoods of millions depend on the ocean and its health; therefore, maritime security must be managed.

The purpose of this chapter is to provide an overview of maritime security in the waters surrounding sub-Saharan Africa. The first section looks at criminal threats common to the maritime region that in turn threaten the achievement of blue economy goals. The second section critically explores regional responses to those maritime crimes. The final section explores how corruption inhibits the effectiveness of those responses. The intended outcomes of this research are therefore of great regional importance, particularly to achieve blue economy goals. Maritime security supports the blue economy goals to reduce poverty, offer new opportunities, protect the region's natural assets, provide resilience to climate change and population growth, and mitigate obstruction of important trade routes operating through the region. Without effective maritime crime prevention architecture to achieve these goals, insurmountable challenges will present themselves in the near and long-term future.

Maritime insecurity explained

Maritime piracy

Low levels of maritime piracy exist around the world, although globally the East and West African regions face the highest level of risk due to the expansive nature of the ocean, limited surveillance and guardianship, and the high numbers of vessels transiting the important trade routes (Brume-Eruagbere, 2017; Lindley, 2016; Onuoha, 2012). Reported attacks show that over the past three decades, despite peaks and troughs, maritime piracy remained a constant feature around the sub-Saharan African coast, with Somali and Nigerian piracy at its worst reportedly experiencing more attacks than the rest of the world's piracy attacks combined (ICC-IMB, 2003, 2009, 2014, 2019, 2020). While the nature and extent of maritime piracy in the region surrounding sub-Saharan Africa varies, two distinct hotspots exist: Somali piracy around the Horn of Africa and Nigerian piracy around the Gulf of Guinea. In both cases perpetrators are highly organized, armed with automatic weapons, and their modus operandi is to take hostages for ransom (ibid., 2020). This has proved to be an effective business model, albeit one that comes at a high economic, social and human cost. The ongoing nature and extent of piracy serves as a useful 'litmus test' for the potential for other maritime crimes proliferating within the region.

Piracy is a clear threat to maritime security and by extension to blue economy goals. Piracy occurs both within international waters (UN, 1982: Article 101) and similarly armed robbery against vessels occurs in a state's EEZ (IMO, 2009a: Annex, para. 2.2). For the purposes of this chapter, 'piracy' refers to activities within and beyond the EEZ. Piracy in the oceans surrounding sub-Saharan Africa exists both within and beyond the EEZ and therefore responses must involve holistic domestic and regional measures.

Illegal fishing

In 2006 illegal fishing caused a loss in profits of more than US $1 billion to sub-Saharan African countries, as well as critically depleting fish stocks (The Brenthurst Foundation, 2010). Fisheries crimes are carried out within a legitimate industry, unlike drug crimes, for example, which are always illegal. Weakened governance surrounding fisheries also exposes the sector to other crimes such as corruption and document fraud, whereby companies may seek favourable treatment from authorities who issue licences and quotas, or who are prepared to overlook breaches (de Coning, 2011). Fishing license fraud has affected littoral sub-Saharan African states such as Nigeria, Somalia and Tanzania (see, for example, Lindley, 2016; Mogotsi, 2019; Stop Illegal Fishing, 2016). These facilitating crimes further undermine regional maritime security. Worse still, the Organisation for Economic Co-operation and Development (OECD) (2013: 43) reported that widespread vulnerabilities to tax crime exist in the fisheries sector, including tax, customs and social security fraud. Furthermore, open registries (or flags of convenience) were found to facilitate the falsification of records, directly benefiting owners and companies (de Coning, 2011; OECD, 2013: 32). Given the extensive nature of fishing off the sub-Saharan African coast, the inability to control illegal fishing directly affects the region's fishing industry and broader community, and therefore its blue economy agenda. Greater political will is needed to achieve SGDs 14.4 and 14.6, and to harness fishing industry opportunities in line with blue economy goals to build a sustainable fishing future.

The target of SDGs 14.4 and 14.6 is to end illegal, unreported and unregulated (IUU) fishing through greater regulatory control (UN, 2015). Several international instruments exist to address IUU fishing in some way. This includes instruments covering forced labour on board vessels, sustainable fisheries and vessel safety, for example. Importantly, IUU fishing is defined in the International Plan of Action to Prevent, Deter and Eliminate Illegal, Unreported and Unregulated Fishing (IPOA-IUU) (FAO, 2001). Lindley and Techera (2017) highlight the complexities of responding to IUU fishing in a matrix combining governing international organizations and their hard and soft instruments. Given the widespread global commitment to counter illegal fishing with the help of a number of international organizations, it may be expected that illegal fishing should have been brought under control by now; but it has not (Telesetsky, 2014). This chapter deals solely with the *illegal* component of IUU fishing.

Trafficking of drugs, humans and other contraband

Transnational organized crime in the region surrounding sub-Saharan Africa tends to involve the marine environment due to the extensive maritime border, which facilitates the undetected entry of criminals to the land. Fishing vessels, yachts and other pleasure craft as well as larger commercial vessels are used as covert drug transporters, human traffickers, migrant and contraband smugglers, and carriers of stowaways (Biegus aqnd Bueger, 2017; Hübschle, 2010; Malcolm and Murday, 2017; Walters, 2008). Each of these examples are gaining priority from law enforcement across sub-Saharan Africa, to varying extents.

Small island states off the coast of Africa are particularly at risk. Organized criminal syndicates may use outlying unpoliced islands as transit points for their illicit activities to avoid detection. For example, the expansive spread of Indian Ocean islands such as Mauritius and Seychelles plays a role in supporting the regional drug trade (UNODC, 2019). While this connection is well known, the nature and extent thereof may be less understood and responses have been limited, or have had limited effects (Malcolm and Murday, 2017). Drug trafficking has emerged as a priority for Mauritius, which focuses on preventing criminal imports of trafficked drugs via a newly established, dedicated policing body (ibid.). Mauritius's proactive approach is promising, because the issue is not isolated and weak law enforcement and/or corruption at the border make it easier for drugs to be brought into the country. The continuation of these illicit activities fails blue economy goals, and SGDs 14 and 16, focused on healthy oceans, and safe and just communities, respectively, within the region.

The sub-Saharan African region is also exposed to human trafficking, facilitated by ocean transportation. The 2019 US Department of State *Trafficking in Persons* report consistently identifies several states within the region as either Tier 2 (such as Angola and South Africa), whereby governments do not fully comply with international counter-trafficking standards, or as Tier 3 (such as Eritrea and Equatorial Guinea), which totally fail to comply with them (United States Department of State, 2019). In sub-Saharan Africa, the nature and extent of vulnerable age groups and genders of trafficked victims is indeed diverse. In Africa children may frequently be trafficked for exploitation in the fishing industry; commonly they come from West Africa and are moved east to the Great Lakes and the African islands (UNODC, 2016), while adults may be trafficked to work on board IUU fishing vessels belonging to several littoral states, such as Seychelles, South Africa, Guinea and Tanzania (de Coning, 2011: 44; United States Department of State, 2019: 413, 428, 451, 453). Since the 1990s the number of irregular migrants, many originating in sub-Saharan Africa, entering Europe from Africa via the Mediterranean has increased due to tougher immigration policies in Europe (UNODC, 2011b: 10, 19). Greater security at ports, through strengthened law enforcement, may increase the likelihood that victims are intercepted at the border.

The lack of border security enables these vulnerable ports to become 'ports of convenience' for various crimes (Petrossian, 2014). For example, vessel stowaways are able to move freely between borders (IMO, 2019). Stowaways crossing borders without the necessary documentation amounts to illegal migration, often interlinked with organized crime. Ports in Durban, South Africa, and Lagos, Nigeria, have the highest incidence of stowaways (ibid.) and in response South Africa has banned entry for all stowaways (P&I Associates, 2019). The facilitation of entry of illegal stowaways by corrupt law enforcement personnel, whose role it is to protect the borders, cannot be overstated.

Porous borders, corruption and weak law enforcement all facilitate other forms of contraband trafficking in the sub-Saharan African region. Contraband may include various drugs, gold, diamonds, charcoal (UNODC, 2011b), wildlife, fuel, cigarettes, alcohol, medicines, electronics and audio-visual goods, timber, firearms, counterfeit currency (Hübschle, 2010) and sugar (Petrich, 2019). Weak governance on land enables these contraband goods to be transported by sea, to the detriment of the blue economy agenda.

Responses to maritime crimes

Maritime security is a domestic concern, although owing to likely inconsistencies between bordering states' responses, regional responses are also needed. These are useful for filling the gaps. International instruments, both binding and non-binding, provide a platform from which domestic and regional responses can be built. However, as is the case with all international instruments, put simply international laws are useful when they have entered into force; when there is unified adoption of such laws across the region; when obligations have been met by each state party to the treaty; and when they are binding and enforceable (Guilfoyle, 2019; Lindley and Techera, 2017; Waldron, 2011; Wessel, 2015). If these conditions are not met, such laws may be unable to achieve their purpose. However, adoption and implementation of obligations set out in existing international laws support regional cooperation to achieve maritime security when coupled with focused multilateral and even bilateral agreements that target region-specific issues and provide tailored methods to respond more effectively. Strategically aligning the region through enforcement capabilities and a web of international/regional instruments and agreements ensures a harmonized response when addressing domestic or regional maritime threats and vulnerabilities. International arrangements responding to common regional threats must be reflected in domestic laws designed to facilitate state-led investigations into criminal and their prosecution.

Given the potential for crimes occurring in the waters surrounding sub-Saharan Africa to spill over into the region, there is sense in having interconnected responses, as an isolated efforts could be financially unsustainable, inconsistent and overlapping, and therefore less effective in the long term. To address regional maritime security threats that emerged over the past decade, a vast toolkit of global and regional responses has become available. Some responses are broad-ranging, while others address specific concerns.

Owing to increased interest in securing its maritime domain and developing its blue economy, the African Union (AU) adopted two soft law – non-legally binding – strategies relevant to maritime security: the 2050 Africa's Integrated Maritime Strategy (2050 AIM Strategy) in 2014 and the African Charter on African Maritime Security and Safety and Development in Africa (or Lomé Charter) in 2016. The 2050 AIM Strategy sets out an optimistic roadmap in line with existing international law and is based on cooperation to address 'insecurity, various forms of illegal trafficking, degradation of the marine environment, falling biodiversity and aggravated effects of climate change' (AU, 2014: 7). With concern for the blue economy at its core, the Lomé Charter focuses on the prevention and control of transnational organized crime at sea, the prevention of accidents at sea and the promotion of sustainable maritime sectors (AU, 2016: Article 4). While the Lomé Charter is a significant achievement (Egede, 2018; Fantaye, 2018), the aspirational framework lacks the critical impetus necessary for it to enter into force (Brits and Nel, 2018).

Research shows that frequently criminals engaging in legal activities in the maritime environment shift their focus to criminal opportunities as they arise. Therefore, responses need to be holistic geographically and by the type of crime so as to cover as many potential crimes as possible. This demonstrates the need to embrace innovative options to ensure agility in responding to threats. For example, regional fisheries management organisations (RFMOs) may usefully be able to guide responses to maritime crimes beyond illegal fishing. RFMOs contribute to IUU fishing vessel blacklists (Trygg Mat Tracking, 2020). It is well known that IUU fishing vessels are used to conduct other crimes, such as drug trafficking and maritime piracy (UNODC, 2011a; de Coning 2011). For example, organized criminals were found to operate legitimate fish processing facilities in Ghana to provide a cover for their illegitimate drug activities and to successfully divert unwanted attention from law enforcement (ibid.: 85).

Maritime piracy

In response to Somali piracy, the International Maritime Organization (IMO), a specialized agency of the UN responsible for regulating shipping, adopted the Djibouti Code of Conduct (Djibouti Code) in 2009 (IMO, 2009b: para. 4). This instrument harmonized the responses of invited East African states (IMO, 2009b). Such states were required to have accepted the UN Security Council resolutions and existing international law in their responses to piracy. Similarly to other UN instruments, the Code requires signatories to align national legislation with international laws on piracy and armed robbery against ships. Significantly, the Djibouti Code outlines a framework for investigation, arrest and prosecution; seizure of suspect ships and property aboard such ships; the rescue of ships, persons and property victimized by Somali piracy; and conduct shared operations with signatory states and with navies from countries outside the region (ibid.: para. 9). In order to support these activities, the Djibouti Code identified the need for unified infrastructure in the region. These long-term solutions for the region seek to ensure the development of cohesive maritime laws and regulations, and to create a platform for positive collaboration within the region.

Responding to the need to broaden its scope to cover other illicit maritime activities, including human trafficking and IUU fishing, the IMO adopted the Jeddah Amendment to the Djibouti Code (IMO, 2017). This helped to enhance interest in protecting the blue economy linked the need to address causes of regional maritime insecurity (ibid.). By 2017 15 of the 17 eligible states had signed the Amendment (ibid.).

The control of Somali piracy through regional cooperation that is central to the Djibouti Code brought to light the need to adopt a response to piracy off the west coast of Africa. As such, the Yaoundé Code of Conduct (Yaoundé Code) emerged in response to piracy off the Gulf of Guinea. It mirrors the Djibouti Code, yet broadens the scope to additionally include transnational organized crime and maritime terrorism (United Nations Security Council, 2012). Similarly to the Djibouti Code and other regional instruments that seek to improve maritime security in order to enhance the blue economy, the Yaoundé Code promotes cooperation (IMO, 2013: Article 1(5)). To achieve shared goals, regional cooperation is required that includes implementing domestic legislation to criminalize piracy and other relevant maritime crimes; sharing information between signatories; repatriating victims of marine crime; and interdicting and prosecuting suspected offenders (ibid.: Article 2(1)). Problematically, there continues to be limited prosecutions in the Gulf of Guinea as regional states continue to work towards adopting relevant anti-piracy legislation (Ogbonnaya, 2020; UNODC, 2018); however, this successful strategy is only one of several important strides towards controlling piracy.

There is a clear impetus for sub-Saharan African states to take a proactive approach to responding to criminal threats, including and especially piracy. The AU's Lomé Charter draws together the Jeddah amendment and Yaoundé Code and encourages states to align their responses, share information and intelligence, and cooperate in responding to maritime threats (AU, 2016: Articles 5–8, 32, 33 and 34). Timely and transparent information and data transfer between stakeholders is believed to be one of the greatest weapons against piracy (IMO, 2017). The lack of urgency afforded to information exchange may inhibit the response to piracy. The approach adopted by the AU is indeed comprehensive and encompasses the blue economy agenda as well as the maritime security threats facing Africa.

Illegal fishing

At the international level, the international plan of action to prevent, deter and eliminate IPOA-IUU is a highly relevant (but soft law) instrument that provides a widely accepted definition of IUU fishing (FAO, 2001). While the IPOA-IUU is primarily concerned with food security and supports the conservation of fish stocks and the sustainable development of fisheries by providing states with a framework to implement regional and domestic responses, it contributes to maritime security (ibid.). The IPOA-IUU requires that vessels are authorized to fish in waters beyond their jurisdiction, but does not specifically address living and working conditions on board vessels, potentially leading to situations of forced labour and human rights abuses (United States Department of State, 2019). Weak law enforcement and corruption increase the likelihood of illegal fishing in the region surrounding sub-Saharan Africa, and while it is a global instrument, it has important regional application.

A regional approach to addressing illegal fishing ensures consistency in responding to criminal threats that weaken maritime security. Unfortunately, the AU's Lomé Charter does not extend to illegal fishing; however, the Yaoundé Code and Jeddah Amendment do include illegal fishing and as such signatories to both instruments must declare their willingness to combat it, along with other marine-based crimes taking place in the region (IMO, 2013). While not binding on signatories, it provides a layer within the maritime security architecture that states can build on. Beyond the Yaoundé Code and Jeddah Amendment, RFMOs have an enveloping geographic and species-specific role that importantly broadens security for the region (Trygg Mat Tracking, 2020). For example, sub-Saharan Africa is bound to RFMOs such as the South-East Atlantic Fisheries Organisation and the Indian Ocean Tuna Commission. Additionally, cooperative initiatives such as FISH-i

Africa pool efforts between participating states to unify the response to IUU fishing off the East African coast (Stop Illegal Fishing, 2020). Regionally generated initiatives may be more targeted and sustainable in the long term.

Trafficking

Human trafficking and modern-day slavery are closely linked to the maritime domain and therefore have an impact on the blue economy. In 2000 the international community adopted the Convention on Transnational Organized Crime and three supplementary protocols, including the Protocol to Prevent, Suppress and Punish Trafficking in Persons, Especially Women and Children (Trafficking Protocol) (UN, 2000). Widespread uptake of the Trafficking Protocol across African states, with exceptions such as Somalia and South Sudan, shows political will in stamping out these practices; however, several areas remain where improvements are possible. For example, it is necessary not only to address the criminal activities of the traffickers, but also factors that increase vulnerability to trafficking recruitment and external enablers, such as corruption at the border (Lindley, 2020).

While there are areas for improvement, some positive gains are notable. For example, the Southern African Development Community Anti-Trafficking in Persons Network links to, and shares data with law enforcement to increase intelligence about victim and trafficker profiles, trafficking routes, traffickers' methods and types of exploitation, assessments of victim services, and the status of investigations and prosecutions (United States Department of State, 2019). This information is integral in preventing future trafficking and strengthening investigation and prosecutorial success, which in turn is likely to act as a deterrent to criminals as the benefits fail to outweigh the risks.

Indeed, trafficking in human beings has a direct link with the marine environment. As such, the UN's Protection at Sea Initiative focuses on protecting people who are vulnerable to illegal movement by sea (UNHCR, 2015). Recognizing the complexities associated with protecting against criminal organization-assisted migration, it therefore encourages cooperation among key stakeholders to share the challenge of securing borders and supporting political and strategic initiatives to enable a regional response (ibid.). Understanding how the vulnerabilities present themselves is important to most appropriately direct the resources and build capacity to respond effectively and sustainably.

The impact of corruption on response effectiveness

While significant progress has been made as a result of the emergence of international and regional organizations and the adoption of instruments that seek to boost security in sub-Saharan Africa's maritime region, rife corruption yields a net shortfall. Corruption is becoming increasingly ubiquitous in connection with maritime insecurity, particularly off the coast of Africa (de Coning and Stølsvik, 2013; Lindley *et al.*, 2018; UNODC, 2018). In 2004 the UN adopted the Convention against Corruption (UNCAC), overcoming inconsistencies in defining corruption and all related aspects (UN, 2004). UNCAC identifies the involvement of corruption in serious crimes and therefore links closely to other relevant UN instruments (ibid.: preamble). Importantly, it has achieved almost universal adoption (UNODC, 2020). UNCAC recognizes its role in supporting signatories to contain corruption (UNICRI, 2019).

Corruption undermines the blue economy. For example, Icelandic fishing and fish processing giant Samherji was at the centre of the Fishrot scandal involving bribes made to Namibian government officials to secure Namibia's coveted fishing quota, who over a four-year period laundered kickbacks valued at N$150 million in tax havens (Mogotsi, 2019). The investigation uncovered a culture of systemic corruption and disregard for sustainable fishing practices, exercised over several years. The offenders faced charges ranging from corruption to money laundering and fraud as well as for contravening the Namibian fisheries laws (Menges, 2019). Corruption among public officials in positions of power, who are entrusted by the people, shows blatant abuse of power for personal gain. Indeed, this is not an isolated example of maritime security-related corruption. In another case, a criminal syndicate relied on corrupt officials to obtain false Tanzanian fishing licenses. Authorities intercepted at least 10 Taiwanese-flagged vessels operating under illegal licenses in Tanzanian waters (Stop Illegal Fishing, 2016). While no offenders involved in the scandal were prosecuted, it acted as a strong deterrent, leading others perpetrators operating under similarly fraudulent licenses to obtain legitimate licenses and made it risky to conduct such corrupt practices (ibid.). The consequences of such corrupt activities may have a catastrophic impact on the regional blue economy.

Conclusion

This chapter illustrated crimes that are common to sub-Saharan Africa's maritime region, namely maritime piracy, illegal fishing and trafficking of drugs, human beings and other contraband. True, there are a number of international binding and non-binding agreements in place to support the control of maritime crimes which threaten the region's security. However, a lack of political will and limited resources overshadow the effectiveness of these instruments. Additionally, there is a strong link between maritime crime and corruption in sub-Saharan Africa. Corruption facilitates the type of crimes explored in the first section of the chapter, along with weak law enforcement to control it at the borders, both on land and at sea. Strengthening maritime security is key to protecting and boosting the blue economy and regional cooperation to enhance responses is critical to their success.

References

African Union (AU) (2014). *2050 Africa's Integrated Maritime (AIM) Strategy.* Addis Ababa: AU. https://au.int/en/documents-38.

African Union (AU) (2016). *African Charter on African Maritime Security and Safety and Development in Africa (Lomé Charter)*. Lomé: AU. https://au.int/sites/default/files/treaties/37286-treaty-0060_-_lome_charter_e.pdf.

Beuger, C. (2015). 'What Is Maritime Security?' *Marine Policy*, 53, 159–164.

Biegus, O., and Bueger, C. (2017). 'Poachers and Pirates: Improving Coordination of the Global Response to Wildlife Crime'. *SA Crime Quarterly*, 60(June).

Bradford, J. F. (2005). 'The Growing Prospects for Maritime Security Cooperation in Southeast Asia'. *Naval War College Review*, 58(3), 63–86.

Brits, P. and Nel, M. (2018). 'African Maritime Security and the Lomé Charter: Reality or Dream?' *African Security Review*, 27(3–4),226–244. doi:10.1080/10246029.2018.1546599.

Brume-Eruagbere, O. C. (2017). *Maritime Law Enforcement in Nigeria: The Challenges of Combatting Piracy and Armed Robbery at Sea* (Master of Science in Maritime Affairs), World Maritime University, Malmo. https://catalog.wmu.se/cgi-bin/koha/opac-detail.pl?biblionumber=79671 (555).

de Coning, E. (2011). *Transnational Organized Crime in the Fishing Industry, Focus on: Trafficking in Persons, Smuggling of Migrants, Illicit Drugs Trafficking*. Vienna: UNODC. www.unodc.org/documents/human-trafficking/Issue_Paper_-_TOC_in_the_Fishing_Industry.pdf.

de Coning, E. and Stølsvik, G. (2013). 'Combating Organised Crime at Sea: What Role for the United Nations Office on Drugs and Crime?' *International Journal of Marine and Coastal Law*, 28. doi:https://doi.org/10.1163/15718085-12341263.

Egede, E. E. (2018). *Maritime Security: Horn of Africa and Implementation of the 2050 AIM Strategy.* Uppsala: Life & Peace Institute. https://works.bepress.com/edwin_egede/27/download/.

Fantaye, D. (2018). *Maritime Insecurity Dilemmas amidst a New Scramble for the Horn?* Uppsala: Life & Peace Institute. https://works.bepress.com/edwin_egede/27/download/.

Food and Agriculture Organization of the United Nations (FAO). (2001). *International Plan of Action to Prevent, Deter and Eliminate Illegal, Unreported and Unregulated Fishing*. Rome: FAO. www.fao.org/3/y1224e/Y1224E.pdf.

Fragile State Index (2019). *Fragility in the World 2019*. https://fragilestatesindex.org/.

Guilfoyle, D. (2019). 'The Rule of Law and Maritime Security: Understanding Lawfare in the South China Sea'. *International Affairs*, 95(5), 999–1017.

Hübschle, A. (2010). *Organised Crime in Southern Africa*. Pretoria: Institute for Security Studies Africa. https://media.africaportal.org/documents/OrgCrimeReviewDec2010.pdf.

International Chamber of Commerce's International Maritime Bureau (ICC-IMB) (2003). *Piracy and Armed Robbery Against Ships: Report for the Period 1 January to 31 December 2002*. London: ICC-IMB.

International Chamber of Commerce's International Maritime Bureau (ICC-IMB) (2009). *Piracy and Armed Robbery Against Ships: Report for the Period 1 January to 31 December 2008*. London: ICC-IMB.

International Chamber of Commerce's International Maritime Bureau (ICC-IMB) (2014). *Piracy and Armed Robbery Against Ships: Report for the Period 1 January to 31 December 2013*. London: ICC-IMB.

International Chamber of Commerce's International Maritime Bureau (ICC-IMB) (2019). *Piracy and Armed Robbery Against Ships: Report for the Period 1 January to 31 December 2018*. London: ICC-IMB.

International Chamber of Commerce's International Maritime Bureau (ICC-IMB) (2020). *Piracy and Armed Robbery Against Ships: Report for the Period 1 January to 31 December 2019*. London: ICC-IMB.

International Maritime Organization (IMO) (2009a). *Code of Practice for the Investigation of the Crimes of Piracy and Armed Robbery against Ships.* London: IMO. www.imo.org/en/OurWork/security/piracyarmedrobbery/guidance/documents/a.1025.pdf.

International Maritime Organization (IMO) (2009b). *Protection of Vital Shipping Lanes: Sub-Regional Meeting to Conclude Agreements on Maritime Security, Piracy and Armed Robbery against Ships for States from the Western Indian Ocean, Gulf of Aden and Red Sea Areas, C 102/14 (3 April 2009)*. London: IMO. www.imo.org/en/OurWork/Security/PIU/Documents/DCoC%20English.pdf.

International Maritime Organization (IMO) (2013). *Code of Conduct Concerning the Repression of Piracy, Armed Robbery against Ships, and Illicit Maritime Activity in West and Central Africa*. London: IMO. www.imo.org/en/OurWork/Security/WestAfrica/Documents/code_of_conduct%20signed%20from%20ECOWAS%20site.pdf.

International Maritime Organization (IMO) (2017). *Revised Code of Conduct Concerning the Repression of Piracy, Armed Robbery Against Ships, and Illicit Maritime Activity in the Western Indian Ocean and the Gulf of Aden Area*. London: IMO. www.imo.org/en/OurWork/Security/PIU/Documents/DCOC%20Jeddah%20Amendment%20English.pdf.

International Maritime Organization (IMO) (2019). *Consideration and Analysis of Reports and Information on Persons Rescued at Sea and Stowaways: Formalities Connected with the Arrival, Stay and Departure of Persons: Stowaways, FAL 43/13 1 February 2019*. London: IMO. www.imo.org/en/OurWork/Facilitation/Stowaways/Documents/FAL%2043-13.pdf.

Klein, N., Mossop, J. and Rothwell, D. R. (2010). 'Australia, New Zealand and Maritime Security'. In Joanna Mossop, Natalie Klein and Donald R. Rothwell (Eds), *Maritime Security: International Law and Policy Perspectives from Australia and New Zealand*. London: Routledge.

Lindley, J. (2016). *Somali Piracy: A Criminological Perspective.* London: Routledge.

Lindley, J. (2020). 'Policing and Prosecution of Human Trafficking'. In Valsamis Mitsilegas (Ed.), *Research Handbook on Transnational Crime.* London: Edward Elgar.

Lindley, J., Percy, S. and Techera, E. (2018). 'Illegal Fishing and Australian Security'. *Australian Journal of International Affairs*, 73(1). doi:doi:10.1080/10357718.2018.1548561.

Lindley, J. and Techera, E. (2017). 'Overcoming Complexity in Illegal, Unregulated and Unreported Fishing to Achieve Effective Regulatory Pluralism'. *Marine Policy*, 81.

Malcolm, J. A. and Murday, L. (2017). 'Small Islands' Understanding of Maritime Security: The Cases of Mauritius and Seychelles'. *Indian Ocean Islands: Geopolitics, Ocean, Environment. Journal of the Indian Ocean Region*, 13(2), 234–256.

Menges, W. (2019). '"Fishrot 6" to Apply for Bail'. *The Namibian*. 29 November. www.namibian.com.na/85962/read/Fishrot-6-to-apply-for-bail.

Mogotsi, K. (2019). 'LPM Weighs In on Fishrot Scandal'. *The Namibian*. 19 November. www.namibian.com.na/85626/read/LPM-weighs-in-on-Fishrot-scandal.

Ogbonnaya, M. (2020). 'Nigeria's Anti-Piracy Law Misses the Mark'. *ISS Today*. https://issafrica.org/iss-today/nigerias-anti-piracy-law-misses-the-mark#:~:text=Nigeria 's%20Suppression%20of%20Piracy%20and,including%20fixed%20and%20floating%20platforms.

Onuoha, F. C. (2012). 'Piracy and Maritime Security in the Gulf of Guinea '. *Al Jazeera*. https://studies.aljazeera.net/en/reports/2012/06/2012612123210113333.html.

Organisation for Economic Co-operation and Development (2013). *Evading the Net: Tax Crime in the Fisheries Sector*. www.oecd.org/ctp/crime/evading-the-net-tax-crime-fisheries-sector.pdf.

P&I Associates (2019). *Stowaways in South African Ports*. www.gard.no/Content/28266179/Stowaways_SouthAfricanPorts.pdf.

Petrich, K. (2019). 'Cows, Charcoal, and Cocaine: Al-Shabaab's Criminal Activities in the Horn of Africa'. *Studies in Conflict and Terrorism*. doi:doi:10.1080/1057610X.2019.1678873.

Petrossian, G. A. (2014). 'Preventing Illegal, Unreported and Unregulated (IUU) Fishing: A Situational Approach. *Biological Conservation*, 189(September), 39–48. doi:doi:10.1016/j.biocon.2014.09.005.

Roach, J. A. (2004). 'Initiatives to Enhance Maritime Security at Sea'. *Marine Policy*, 28, 41–66.

Secretary-General of the United Nations (2008). *Report of the Secretary-General on Oceans and the Law of the Sea (A/63/63)*. New York: United Nations.

Stop Illegal Fishing (2016). *FISH-i Africa: Issues/Investigations/Impacts*. Gaborone: Stop Illegal Fishing. https://stopillegalfishing.com/wp-content/uploads/2016/07/FISH-i-Africa-Issues-Investigations-and-Impacts_report_WEB.pdf.

Stop Illegal Fishing (2020). *FISH-i Africa*. Gaborone: Stop Illegal Fishing. https://stopillegalfishing.com/initiatives/fish-i-africa/.

Telesetsky, A. (2014). 'Laundering Fish in the Global Undercurrents: Illegal, Unreported, and Unregulated Fishing and Transnational Organized Crime'. *Ecology Law Quarterly*, 41(4), 939–998.

The Brenthurst Foundation (2010). *Maritime Development in Africa: An Independent Specialist's Framework*. Discussion Paper 2010/03. http://uscdn.creamermedia.co.za/assets/articles/attachments/28769_brenthurst_paper_2010_03.pdf.

Trygg Mat Tracking (2020). *Combined IUU Vessel List*. https://iuu-vessels.org/.

United Nations (UN) (1982). *Convention on the Law of the Sea*. New York: UN.

United Nations (UN) (2000). *Protocol to Prevent, Suppress and Punish Trafficking in Persons, Especially Women and Children, supplementing the United Nations Convention against Transnational Organized Crime*. New York: UN. www.unodc.org/documents/middleeastandnorthafrica/organised-crime/UNITED_NATIONS_CONVENTION_AGAINST_TRANSNATIONAL_ORGANIZED_CRIME_AND_THE_PROTOCOLS_THERETO.pdf.

United Nations (UN) (2004). *Convention Against Corruption*. New York: UN www.unodc.org/documents/treaties/UNCAC/Publications/Convention/08-50026_E.pdf.

United Nations (UN) (2015). *Transforming our World: The 2030 Agenda for Sustainable Development A/RES/70/1*. New York: UN. www.un.org/ga/search/view_doc.asp?symbol=A/RES/70/1&Lang=E.

United Nations Economic and Social Council (ECOSOC) (2018). *24th ICE of Southern Africa: Blue Economy, Inclusive Industrialization and Economic*

Development, E/ECA-SA/ICE.XXIV/2018/Info.1, Balaclava, Mauritius, 18–21 September 2018. New York: ECOSOC. www.tralac.org/news/article/13347-24th-ice-of-southern-africa-blue-economy-inclusive-industrialization-and-economic-development.html.

United Nations High Commissioner for Refugees (UNHCR) (2015). *Strategy and Regional Plan of Action - Smuggling and Trafficking from the East and Horn of Africa: Progress Report*. Geneva: UNHCR.

United Nations Interregional Crime and Justice Research Institute (UNICRI) (2019). *Policy Toolkit on The Hague Good Practices on the Nexus between Transnational Organized Crime and Terrorism*. Turin: UNICRI.

United Nations Office on Drugs and Crime (UNODC) (2011a). *Fisheries Crime*. Vienna: UNODC. www.unodc.org/documents/about-unodc/Campaigns/Fisheries/focus_sheet_PRINT.pdf.

United Nations Office on Drugs and Crime (UNODC) (2011b). *The role of organized crime in the smuggling of migrants from West Africa to the European Union*. Vienna: UNODC. https://emnbelgium.be/sites/default/files/publications/report_som_west_africa_eu.pdf.

United Nations Office on Drugs and Crim (UNODC) (2016). *Global Report on Trafficking in Persons 2016*. Vienna: UNODC. www.unodc.org/documents/data-and-analysis/glotip/2016_Global_Report_on_Trafficking_in_Persons.pdf.

United Nations Office on Drugs and Crime (UNODC) (2018). *Global Maritime Crime Programme Annual Report 2018*. Vienna: UNODC. www.unodc.org/documents/Maritime_crime/20190131_-_GMCP_Annual_Report_2018.pdf.

United Nations Office on Drugs and Crime (UNODC) (2019). *World Drug Report*. New York: UNODC. https://wdr.unodc.org/wdr2019/en/prevalence_map.html.

United Nations Office on Drugs and Crime (UNODC) (2020). *UN Convention against Corruption: Signature and Ratification Status*. 6 February. New York: UNODC. www.unodc.org/unodc/en/corruption/ratification-status.html.

United Nations Security Council (2012). *Resolution 2039 (2012)*. New York: UN. http://unscr.com/en/resolutions/doc/2039.

United States Department of State (2019). *Trafficking in Persons Report: June 2019*. Washington, DC: United States Department of State. www.state.gov/wp-content/uploads/2019/06/2019-Trafficking-in-Persons-Report.pdf.

Waldron, J. (2011). 'Are Sovereigns Entitled to the Benefits of the International Rule of Law?' *European Journal of International Law*, 2, 315–343.

Walters, W. (2008). 'Bordering the Sea: Shipping Industries and the Policing of Stowaways'. *borderlands*, 7(3).

Wessel, R. A. (2015). 'Revealing the Publicness of International Law'. In C. Ryngaert, E. J. Molenaar and S. M. Nouwen (Eds), *What's Wrong with International Law?* 449–466. Leiden: Brill.

6 Ports, shipping and transportation

Vivian Louis Forbes, PhD

Introduction

Until quite recently world cargo movements have been increasing, illustrated by an upswing in the world economy since 2017. In 2019 global maritime trade gathered impetus and expanded by 4%, the fastest rate of growth since 2015, thus raising optimism within the shipping trade. By 2019 the total volume of seaborne trade worldwide had reached over 11 billion metric tons, reflecting an additional 411 million tons since 2015, nearly half of which comprised dry bulk commodities.[1] Nonetheless, the coronavirus (COVID-19) pandemic is expected to cause a severe decline in the world's maritime commerce, and sub-Saharan Africa will not be immune to the downturn.

In 2019 the sub-Saharan African countries took centre stage for global terminal operators who made investment commitments which the Organisation for Economic Co-operation and Development (OECD), the European Union (EU) and the United Nations (UN) estimated to have grown by up to 5% during 2019. This projected growth came as a result of improved domestic demand, a better regional business environment, investment in infrastructure and growing trade and investment ties with other emerging economies.[2]

Intra-regional and international maritime trade focuses on port infrastructure and shipping and the associated challenges that for many decades have stymied sub-Saharan Africa's port development and hindered economic growth that relies on maritime routes, road and rail transportation linking the hinterland to the ports. Nonetheless, opportunities exist to take full advantage of the benefits that could accrue from the blue economy if appropriate actions – by the public and private sectors—are taken. For example, according to a report published in late 2019, Nigeria's national maritime authority should look inwards; develop its capacity; expand its infrastructure support; and secure its supply chain in order to enhance its reputation in the comity of maritime nations.[3]

Background

Covering an area of over 35 million sq km, sub-Saharan Africa has a relatively large domestic market that possesses significant opportunities – and challenges – to enhance its blue economy. At early 2020 the population of sub-Saharan Africa stood at 1.1 billion, or about 14.1% of the world's total. With an annual average population growth rate of 2.7%, it is estimated that the region will account for more than half of the growth in the world's population between 2020 and 2050.[4] The UN predicts that the region's population will increase to 2 billion by 2050.[5] This population growth will require increased exports and imports of merchandise and other commodities for a secure and sustainable future. There are significant economic development and trade gaps among the regional countries as well as between them and the developed countries.

The Abuja Declaration signed by members of the African Union (AU) in September 2019 highlighted the shortcomings of marine security in the Gulf of Guinea relating to ocean governance and law enforcement on land and at sea and suggested that concrete actions had to be taken if the region's coastal states wanted to benefit from the blue economy. The call was reinforced during a Global Conference held in Nigeria in October 2019. Greater transparency could also help to improve relationships between government authorities, maritime industry and security agencies in the region. Lack of trust and limited cooperation have often hindered thorough investigations into marine malpractice and hence progress in resolving issues. There is a lack of competition: for example, both Kenya and Tanzania are endeavouring to gain exclusive rights to transport minerals and manufactured goods from their northern and western neighbours to either Mombasa or Tanga.[6] Likewise, stiff competition to attract maritime business is evident around the ports in the vicinity of the Gulf of Guinea. Businesses operating in sub-Saharan Africa face fragmentation in its many forms; for example, it costs less to ship a car from Paris, France, to Lagos, Nigeria, than from Accra, Ghana to Lagos. According to Nigeria's former president Olusegun Obasa njo, 'It is easier to move goods from African countries to Europe and the rest of the world than to trade between one African country and another'.[7]

A report published by the World Bank in 2020 observed that the region remains one of the weakest performing regions in terms of ease of doing business. For example, in Côte d'Ivoire and Cameroon it can take over eight days to process the relevant import and export shipping procedures for maritime transport services. Ports in sub-Saharan Africa are the least efficient of any region. Mauritius (ranked 13th) and Rwanda (38th) are the only economies among the top 50 in the World Bank's Ease of Doing Business Index. Other large economies in the region are Kenya (56th), South Africa (84th), and Nigeria (131st). Meanwhile, South Sudan (185th), Eritrea (189th) and Somalia (190th) are the lowest ranked economies in the region.[8]

A report published by the UN Conference on Trade and Development (UNCTAD) suggested that the region faces seven 'binding constraints'. They are as follows:

> Actual and perceived port inefficiency depicted by longer container dwell time, delays in vessel traffic clearance, lengthy documentation processing,

and lesser container per crane hour are issues that are identified as critical. Statistics infer that nearly 70% of delays in cargo delivery comprise of time spent within African ports;

Inefficient rail and road network in the form of low speed and unreliability (railways), insecurity, congestion, delays at checkpoints, diversions due to frequent machinery maintenance;

Inadequate volume of cargo to permit full capacity utilization of maritime transportation along with its inter-linked modes;

High transportation costs which, on a global average, is between 40% and 60% higher;

Non-compatibility of national transportation development plans with regional agreements on the need for harmonization;

Economic viability and actual loss making in other modes especially in the railways; and,

Inadequacy in human resources and Integrated Computer and Technology Systems to support an efficient and integrated and effective integrated Port Management Information Systems (PMIS) to guarantee globally competitive and high-quality port performance.[9]

The pattern of ownership of vessels varies widely between the registries, as will be discussed below. There are four large registries, which collectively share 50% of world gross tonnage by registry (according to 2013 estimates). Owners from five countries together account for nearly 53% of the world tonnage. Among the top 35 ship-owning countries and territories, 17 are in Asia, 14 in Europe and four in the Americas.[10]

Shipping movements and trends

International trade can be broadly classified as trade in consumer and producer goods and also in services, such as that provided by cruise and passenger ships. National, regional and global shipping movements generally mirror the world's economy. The tracking of various types of vessel based on their purpose, size and type of cargo showed that 7,999 ships (equating to 3.7% of the world's fleet) sailed into East African ports during 2018. A slightly higher number of ships (9,120 or about 4.3% of the world's fleet) were bound for ports in South Africa. A greater number of ships (12,364 or about 5.8% of the world's fleet) were carrying cargo bound for ports along the West African coast. Sub-Saharan Africa's overseas trade growth rate in 2018 to other regions was to North America (5%), Latin America (11.4%), Asia (15.2%), European countries (15.2%) and Oceania (8.6%). These figures illustrate the importance of trade between African countries and other regions. An indication of the amount of cargo transiting the region's ports showed gross tonnage and the percentage of world trade that it represented in 2018, namely to Eastern Africa (434,999,000 metric tons or 6.0%),

Western Africa (406,765,000 tons or 5.6%) and Southern Africa (475,974,000 tons or 6.6%).[11]

Sub-Saharan Africa relies on its ports, rail and road transportation and maritime trade employing ships to service the 24 coastal and 16 landlocked countries. For example, during 2017 nearly 40% of all merchandise exported by sea comprised crude oil, while nearly 20% of imports comprised petroleum products and gas. Over 66% of imports were categorized as 'dry cargoes', namely dry bulk and containerized goods.[12]

Modes of shipping can be characterized by cargo type: 'container' cargo is measured in twenty-foot equivalent units (TEUs) and such containers have revolutionized shipping during the past three decades. Containers can carry processed, semi-processed and unprocessed goods and they increase efficiency via intermodal systems (for example, rail to port directly to and from the ships); 'break bulk cargo' comprises other dry cargo (for example bauxite/alumina) and main bulk (which includes iron ore, grain and coal); 'tanker trade' refers to the carriage by ships of crude oil, refined petroleum products, gas and chemicals; and 'roll-on/roll-off' vessels are generally used to transport automobiles by sea. The distribution routes and patterns of maritime trade routes are varied. For example, the container ship routes are concentrated in the eastern/western belt around the southern part of the northern hemisphere.[13]

The five main bulk dry cargoes (iron ore, coal, grain, bauxite/alumina and phosphate rock) and oil and gas trade are focused on the sources of these cargoes. Ships' itineraries are generally affected by changes in the volatile world market prices for these commodities. The carriage of bulk dry cargoes and oil and gas tends to have a higher proportion of return voyages in ballast since these ships usually return to the region where the hydrocarbon is sourced – in the Persian Gulf, for example. The mineral and other commodity cargoes are primarily found on routes originating from ports in sub-Saharan Africa, South America and Australia and heading to East Asia (primarily the People's Republic of China), Europe, Canada and United States. Significant changes in maritime traffic routes could result from increases in the extraction of hydrocarbons from the land and offshore reserves of the region (see Chapter 4 in this volume).

Sub-Saharan Africa's maritime trade

The growth rate of sub-Saharan Africa's export volume has trailed behind imports since 2005 partly due to a higher concentration of exports (generally primary products and natural resources). As the terms of trade began to deteriorate from 2012 onwards, diversifying the region's exports became more imperative than ever. Since 2015 the volume of imports has quadrupled while exports have doubled. In 2018 the collective volume of cargo loaded at the region's ports was over 11 million metric tons, comprising 3,194 million tons of tanker trade, 3,210 million tons of main bulk and 4,601million tons of other dry cargo.[14]

Nearly 90% of the region's imports and exports are transported by sea. The AU referred to the shipping industry as a new frontier for the continent's renaissance. However, despite significant opportunities for expansion, the region still makes up just 2.7% of global trade by value, although it does contribute a higher share to global seaborne trade – 7% of maritime exports and 5% of imports by volume.[15]

In 2012 Sub-Saharan Africa's exports of merchandise goods amounted to US $625 billion; however, by 2016 exports had decreased to $361 billion. Exports of services amounted to $106 billion in 2014 but declined to a mere $96 billion in 2016. African countries' exports started decreasing as early as 2013 and the decline accelerated in 2015 to contract by about 29% per annum, similarly to falls witnessed during the 2009 Great Recession. The fall was mainly due to declines in export commodity prices mainly as a result of decreased demand from China that led to a deterioration in terms of trade for many sub-Saharan African countries.

Many countries in the region are small, open economies. They achieve high openness rates –exports plus imports relative to gross domestic product (GDP) – due to their small resource endowments. These countries rely heavily on imports as they produce so few manufactured consumer goods. For example, Seychelles and the Democratic Republic of the Congo import up to nearly 100% of their consumer goods, while Nigeria imports well below the world average of 53.4%.

The region must now take advantage of the economic potential of its ports and shipping sector if it is to realize its growth ambitions. A 2018 study showed that a 25% improvement in port performance could increase GDP by 2%. It inferred that more needs to be done at five of the region's ports that are vying to become regional shipping powerhouses. Transforming these ports into more efficient hubs will not only reduce the high port rates but will also contribute towards inter-African economic integration as well as the region's fuller integration into the world trading system.[16]

Intra-regional and international maritime trade

From 2005–16 the region's trade increased by an average annual rate of 12.5%. As discussed above, most of the region's exports are primary commodities, and Asia, the United States and Europe remain the region's main trading partners. Intra-regional trade is low (perhaps 15% of total trade). The intra-regional trade growth rate in 2016 was about 12.5%.[17]

Maritime transport remains the region's main gateway to global markets. However, problems persist at many of the region's ports. These range from capacity issues to inefficient handling times, customs clearance and documentation checking (in general electronic data interchange, for example, ship's manifest, invoices, bills of lading, certificates of insurance, as well as other documentation required under documentary credits, payment processing arrangements and customs clearance), inadequate security, in many places corruption, poor infrastructure facilities and rail connections, as well as terrible road conditions resulting in generally high transport costs.[18]

Indigenous shipping enterprises

There are few indigenous shipping lines in sub-Saharan Africa. The Liberian Ship Registry, having served the shipping industry since 1948, is the fastest growing major open system operating the 'flag of convenience' concept. Shipowners rely on countries that offer a flag of convenience as it gives merchant mariners the freedom to employ cheap labour from the global labour market as a means of economizing. Seafarers increasingly work on ships flying the flag of a country other than their own or that of their employers. Among the advantages to foreign owners of Liberian-flagged ships are tax advantages; they do not need to be incorporated in Liberia; or carry Liberian seafarers, and there are no provisions for Liberian inspection or control of any ship registered under that flag. There are in excess of 4,400 vessels in Liberia's fleet equating to 12% of the world's ocean-going fleet and over 170 million gross tons. The Liberian Ship Registry is a sovereign maritime jurisdiction responsible for the registration, regulatory enforcement and safety of its ocean-going ships. The registry establishes identification details for ships and records legally enforceable documents, such as mortgages and bills of sale.[19]

The Ethiopian Shipping and Logistics Services Enterprises (ESLSE), established in 1964, known commercially as Ethiopian Shipping Lines, is the national cargo shipping company of Ethiopia. Its main base is the port in Djibouti as Ethiopia lacks a seacoast (it became landlocked in 1993 when Eritrea opted for independence from the Eritrean-Ethiopian Federation that was formed in 1952). The ESLSE focuses its operations on the designated export and import routes to promote foreign trade and generally has been profitable.[20]

Infrastructure issues and customs clearance

Port infrastructure, ships and logistics support are important developments towards enhancing the region's blue economy. Although an efficient and low-cost transport system will not guarantee export success, it is a precondition for most of the region's countries to achieve competitiveness in the global trade. Governments must make significant investments in infrastructure, such as building good roads to the port and providing truck parks. Failure to undertake these tasks will only prolong the port gridlocks.

There is increasing evidence that enhanced transport time decreases trade volumes and increases logistics costs, notably because of increased inventories. Increased transport time dramatically reduces trade efficiency and profitability. Reducing inventories is an important objective of most manufacturers (with 'just-in-time manufacturing'). Without reduced cargo dwell time predictability, the objective of reducing inventories is not likely to be met. Long cargo dwell times in ports are a critical issue in sub-Saharan Africa. Overland transport to landlocked countries in the region consumes over 50% of total transport time from ports to hinterland cities. When it comes to trading across borders and paying taxes, it takes businesses in the region about four days to comply with documentary requirements for processing imports and exports, the longest such delay in the world.[21]

The World Bank's Logistics Performance Index (LPI)[22] assesses 160 countries' logistics efficiency based on their customs clearance processes, the quality of trade and transport-related infrastructure, the ease of arranging competitively priced shipments, their ability to track and trace consignments, and the frequency with which shipments reach the consignee within the schedule time. Its 2018 study analysed 36 countries in the region. South Africa had the best regional ranking in 36th position, the only African country in the top 50, with Côte d'Ivoire next in 43rd position. Angola ranked 162nd, the world's second lowest. By comparison, Germany ranked first.

Customs clearance – a major process in cargo handling at any port or airport – can make a significant contribution to the overall poor performance of port efficiency. This indicator scores consistently lower than the overall LPI of most countries surveyed. In this context, even though South Africa ranked 33rd in the world, it outperformed all the other regional ports. Customs clearing processes remain tedious owing to governments' refusal to invest in customs automation and to acquire scanning machines, particularly in Kenya and Nigeria. The average dwell time for cargo in selected sub-Saharan African ports are Durban, South Africa (14 days), Douala, Nigeria (19), Lomé, Togo (18), Tema, Ghana (20), Mombasa, Kenya (11) and Dar es Salaam, Tanzania (14). Efficient cargo delivery to and from the various ports in these countries remains a Herculean task owing to governments' refusal to repair and upgrade the port access roads.[23]

There are also concerns that ships are being forced to queue at sea before off-loading due to space constraints within the port terminals, and as a result cargo delivery times have slowed down considerably. Importers and clearing agents complain about delays in moving their cargoes in and out of the Apapa and Tin Can Island ports in Nigeria, because the traffic gridlock has worsened dramatically, with large volumes of cargo trapped at the ports. The gridlock at Apapa costs the country US $19 billion annually – a loss higher than the country's entire 2016 budget. A 'massive systems failure' and severe traffic congestion in Mombasa in December 2019 paralysed the port's activities, thereby highlighting some of the problems facing Kenya's maritime strategy.[24]

The US government's Customs-Trade Partnership Against Terrorism (CTPAT) examines the actual and potential threats of terrorist-related incidents at sea and on land. Port operators and shipping are mandated to obtain certification and achieve supply chain security certification, especially for ships destined for ports in the United States increased the problems of documentation such as electronic data interchange, digitalization, inspection and clearance of cargo handling and shipping.[25]

Ports: successes and failures

South Africa's Port of Durban is the largest and generally the best performer in sub-Saharan Africa. South Africa's six ports, plus Douala (Nigeria), Mombasa (Kenya), Nacala (Mozambique), Onne (Nigeria) and Tema (Ghana) along with the new port near Lamu (Kenya, as discussed below) are considered to be hub-attractive in the context of the region.

According to the Transnet National Ports Authority Cargo, a South African government organization, South Africa's Port of Durban catered for 9,821 vessels in 2017 and processed over 22 million metric tons of cargo. However, it is one of the most expensive ports in the world and only achieves 75% of the efficiency expected from a major global hub, lagging far behind other world hubs such as Rotterdam and Antwerp in the Netherlands. In 2018 the latter handled 5.57 million TEUs and the former 7 million TEUs, whereas Durban handled 2.95 million TEUs. Although the 9.5% increase recorded by Durban was impressive the port suffers from crippling congestion with drivers regularly waiting for hours and sometimes days to access the port.[26] Port Elizabeth was rewarded for maintaining high safety standards for its staff, ships and cargo, and is ranked second out of all eight ports in South Africa. The ports of South Africa may be considered exceptional in the context of this discussion. South Africa remains the undisputed port leader of the region, with Cape Town and Ngqura ranking just behind Durban.[27]

The East African market is the least developed in respect of hub port development, although Kenya has positioned Mombasa as a hub port for East and Central Africa. However, it is under-performing in the context of the entire East African region despite its improved performance over the past two decades. There is optimism that Mombasa will emerge as a hub port, but it will require the Kenyan government to increase investments into the port, while the private sector will need to provide efficient operations. The Kenyan government is now planning a huge new port at Lamu, while Tanzania is developing Bagamoyo. Both ports will be larger than any other port in sub-Saharan Africa if completed as planned. They will also be at the centre of much bigger related downstream developments, with industrial zones and intensive farming being proposed.

The Tanzanian authorities hope that Bagamoyo will handle 20 million containers a year, which is 25 times larger than the port at Dar es Salaam. Kenya's planned Lamu port is expected to be just as big. However, it should be noted that these are the proposed long-term figures, which will be achieved over decades rather than years. Construction will take place in phases as and when required.

Construction work on Bagamoyo began at the end of 2019, after financing was granted by China Merchant Holding International and the Oman State General Reserve Fund.. Funds are also being injected into the Tanzanian port of Tanga. The Tanzanian government managed to persuade its counterpart in Uganda to route its planned oil export line through Tanzania to Tanga, rather than through Lamu.. Mombasa will be confronted with stiff competition from Dar es Salaam which received a substantial grant and credits from the World Bank for a Maritime Gateway Project in 2019. The Port of Nacala, Mozambique, is attractive as it has connectivity to its landlocked neighbours Malawi, Zambia and Zimbabwe.[28]

Djibouti plays a key role in the volatile Horn of Africa and the geographic constriction known as the Straits of Bab-al-Mandeb (at the southern entrance of the Red Sea). The port is host to Djibouti's Western allies as well as to China's People's Liberation Army (it acts as a naval base), allegedly for the purpose of assisting as a countermeasure to acts of armed robbery, piracy and terrorism. In 2009 the Doraleh Container Terminal (DCT) was built, funded by DP World, and has since been a major catalyst for growth. While Djibouti's old port lacked capacity and water depth for larger vessels, the DCT can now handle 1.6 million TEUs and has a water depth of 18 m. Although the port serves the only rail link to Ethiopia, it is less likely to become a hub than Kenya and Tanzania due to its strategic geopolitical position and hosting of numerous foreign military bases.[29]

A US $269 million upgrade of the Berbera port container terminal was proposed after the Somaliland Port Authority announced expansion operations. The redevelopment including expansion of the terminal yard is expected to be completed by 2026. The Somaliland Port Authority is a 30% stakeholder in the terminal with DP World.[30]

The Port of Abidjan is the economic hub of Côte d'Ivoire, generating around 80% of the nation's customs revenues. It is also one of West Africa's leading commercial ports, serving landlocked countries including Mali, Burkina Faso and Niger. An expansion project in Abidjan was launched in late 2015 to widen and deepen the canal leading to the main port. Work is also underway to build a second container terminal to increase the port's cargo capacity and host larger, new-generation ships. The project is expected to be completed in 2021 and will cost the authorities around US $1.2 billion. The investment should pay dividends and position Abidjan among the top three African ports likely to emerge as hubs in the coming years.[31]

Nigeria's primary seaport is the Port of Lagos. However, like all ports in the country, it is generally acknowledged that it faces problems with efficiency. In October 2016 Lagos Port published an analysis of the country's ports which claimed that revenue is lost every year due to inefficiencies and infrastructural shortcomings. The huge infrastructure deficit has led to deplorable access roads, faulty cargo scanners, non-existent rail systems and non-functional truck bays which conspire to have a negative impact on service delivery efficiency.[32]

Foreign operators

Given the lack of domestic investment resources it is understandable that sub-Saharan African countries look to outsiders to upgrade their ports. One of the main regional partners is Dubai Ports International (DPI), founded in 1999. DPI expanded rapidly through the acquisition of ports outside the Persian Gulf. Following a merger with Dubai Ports Authority in 2005 DPI changed its name to DP World and in the following year it purchased the fourth largest port operator in the world – the Peninsular and Oriental Navigation Company (P&O) for about US $7 billion. DP World has a portfolio of 78 operating marine and inland terminals supported by over 50 related businesses in 40 countries across six continents. It has operations in Djibouti (two container terminals), Mozambique (three container terminals), Senegal (one container terminal), South Africa (four non-container terminals) and Senegal, at Port du Fulur.[33]

In 2019 DP World was granted a concession to develop and operate a logistics complex in landlocked Rwanda. The 90,000 sq m complex was built on a greenfield site at the capital Kigali and it offers warehouse capacity for 640,000 metric ton of cargo and handles up to 50,000 TEUs per annum. In the same year DP World noted with caution that depending on demand growth, the initial project investment may be expanded. The Shippers Council of Eastern Africa applauded the announcement.[34]

Dar es Salaam's container terminal is operated by Tanzania International Container Terminal Services, an affiliate of Hong Kong's Hutchison Whampoa. The Bagamoyo terminal will be operated by China Merchant Holdings. Lamu is being developed by the China Communications Construction Company but the Kenya Ports Authority will still be in operational control. China's trade recession since 2017 has cast shadows across the global maritime and trade sector. News of multiple investments, during November 2019, in African ports and supply chain infrastructure points to one place where industry leaders are clearly turning for growth opportunities. The total value of Chinese investments and construction in Africa since 2015 is over US $2 trillion. The financing of major projects in sub-Saharan Africa's ports and maritime trade poses a risk that could prevent other foreign companies from competing on subsequent commercial opportunities, essentially imposing restrictions on African governments to transact only with Chinese entities.[35]

Inland transportation

In many sub-Saharan African countries the traffic situation in port vicinities is unbearable. This is particularly evident in Nigeria's Port of Apapa and Kenya's Port of Mombasa. Apart from the deplorable state of the port access road, the traffic gridlock, containers fallen into an inland waterway and sudden rise in haulage cost by truck owners who blame the gridlock on the dilapidated road infrastructure as experienced in Apapa in December 2019. It is difficult to tell if Nigeria made any appreciable progress in 2019. It is even much more problematic to discuss the fate of the maritime industry in future years as it appears that Nigeria's maritime industry is doomed to constant retraction while its politicians fail to act. The Nigerian government appears content with taking the profits from the maritime sector without investing in return. This profit-taking disposition to the sector has left it in desperate need of investment infusions.[36]

Sudden increases in haulage costs is a major issue in Africa because transportation charges depend on distance, location and, importantly, the power of bargaining from on sides – the trucking companies and the sellers and buyers of the cargo that is taken from the ship and/or warehouse to the factory or marketplace.[37]

In 2021 the Tanzanian government awarded a contract to a partnership of China Civil Engineering Construction and China Railway Construction Corporation for the construction of the 341-km Mwanza–Isaka standard-gauge railway that will run parallel to the pipeline connecting the port of Dar es Salaam to the ferry services in operation on Lake Victoria. The Chinese consortium won the contract because it was offering the cheapest option and, probably, also because any line to Lamu, Kenya, would be vulnerable to attack from the Somalia-based Islamist al-Shabaab militant grouping. Unlike Tanzania, Kenya has stuck with state ownership. The Tanzania-Zambia Railway was built in the 1970s by the Chinese government to help Zambia to export its copper through Dar es Salaam. New railways running from Lamu to South Sudan and Ethiopia are planned, while funding is currently being sought for a new line from Rwanda and Burundi to either Dar es Salaam or Bagamoyo.[38]

Conclusion

Sub-Saharan Africa will not escape the ravages of the coronavirus (COVID-19) pandemic and maritime commerce will suffer significant downturns in the near term. In the longer term, there are general concerns about doing business and the efficiencies and robustness of supply chains in sub-Saharan Africa because of a lack of capacity and alternatives. Poor infrastructure makes the situation more difficult especially in landlocked countries where there are serious network and bottleneck problems. The region's ports would benefit substantially from better coordination of infrastructure development, transparency in administration, clearly stated haulage charges and zero tolerance of bribery and corruption. The region must now take advantage of the economic potential of its ports and shipping sector and connectivity if it is to realize its blue economy growth ambitions.

If the region is to fulfil its potential, its ports need to be developed to accommodate larger vessels and to ensure the availability of adequate cargo handling equipment. There needs to be improved access to the hinterland and landlocked countries by multimodal transport and transport corridor approaches through the building of inland transport infrastructure. The regional states must make use of the relevant technologies such as digital solutions to facilitate transport and trade, eliminate inefficiencies, improve processes and enhance transparency. Importantly, it is necessary to enhance transport connectivity either by sea or landwards to reduce transport costs and improve Africa's position in the global shipping networks. The political will must include enabling soft infrastructure in support of transit transport and trade facilitation by harmonizing transport regulation, including road and rail transport. Finally, the sub-Saharan African countries should be encouraged to promote maritime clusters where shipping and port activities can both help to boost related service sectors.

Notes

1 UN Conference on Trade and Development (UNCTAD) (2019) 'Review of Maritime Transport', https://unctad.org/system/files/official-document/rmt2019_en.pdf. Gaël Raballand, Salim Refas, Monica Beuran and Gozde Isik (2018) 'Maritime Transport in Africa: Challenges, Opportunities, and an Agenda for the Future Research. Why Does Cargo Spend Weeks in Sub-Saharan Ports? Lessons from Six Countries', Washington, DC: World Bank. Available at: https://unctad.org/system/files/non-official-document/dtltlbts-AhEM2018d3_WorldBank_en.pdf.

2 Odongo Kodongo and Kalu Ojah (2016) 'Does Infrastructure Really Explain Economic Growth in Sub-Saharan Africa?' *Review of Development Finance*, 6 (2), pp. 105–25; Anton Eberhard, Vivien Foster, Cecilia Briceno-Garmendia, Fatimata Ouedrogo, Daniel Camos and Maria Shkaratan (2008) '*Africa Infrastructure Country Diannostic*, The World Bank and Sub-Saharan Transport Policy Program (SSTPP), 2008; John Bruce Thompson (2011) *Africa's Transport: A Promising Future*, July, World Bank and Sub-Saharan Transport Policy Program, www.ssatp.org/sites/ssatp/files/publication/SSATP-Advocacy-Paper-Jul2011_Combined.pdf, World Bank and SSATP (2008) '*Africa Infrastructure Country Diagnostic*', *Roads in Sub-Saharan Africa, Summary of Background Paper 14* June, World Bank and SSATP, http://documents1.worldbank.org/curated/en/938171468194345853/pdf/551050WP0p109210Box349439B. The full report, with detailed country annexes and technical notes, is available at www.infrastructureafrica.org.

3 'Ship Owners' Move to Harness $3.5bn Nigerian Content in Oil Industry', *International Shipping News*, 27 August 2019, www.hellenicshippingnews.com/ship-owners-move-to-harness-3-5bn-nigerian-content-in-oil-industry/; Ndikom Obed, E. C. Buhari, S. O. Okeke and Matthew W. Samuel (2017) 'Critical Assessment of Maritime Industry in Nigeria: Challenges and Prospects of Policy Issues', 20(3), www.ajpssi.org/index.php/ajpssi/article/view/263.

4 PwC (2018) 'Geographical Classification of S-SA States'; United Nations *World Population Prospects*, 2019, data booklet, www.un.org/development/desa/pd/content/world-population-prospects-2019-data-booklet.

5 Ibid.

6 African Union, *The Abuja Declaration*, September 2019, https://au.int/sites/default/files/treaties/7797-treaty-0041_revised_african_maritime_transport_charter_e.pdf.

7 Samson Echenim (2019) 'Nigeria's Maritime Records Poor Performance in 2019, Edges Up for 2020', 30 December, Business a.m., www.businessamlive.com/nigeria-maritime-records-poor-performance-2019-2020-report/; Chioma Jaja (2011) 'Freight Traffic at Nigerian Seaports', *Social Sciences* 6(4), pp. 250–58.

8 World Bank (2020) *Doing Business 2020 Region Profile of Sub-Saharan Africa*, Washington, DC: World Bank; World Bank (2020) *Doing Business 2020: Comparing Business Regulation in 190 Economies*, Washington, DC: World Bank, http://documents1.worldbank.org/curated/en/688761571934946384/pdf/Doing-Business-2020-Comparing-Business-Regulation-in-190-Economies.pdf.

9 UNCTAD (2013) *Economic Development in Africa Report 2013*, 'Inter-African Trade: Unlocking Private Sector Dynamism', Geneva: United Nations, p. 158, https://unctad.org/system/files/official-document/aldcafrica2013_en.pdf;
UNCTAD (2015) *Economic Development in Africa Report 2015*, New York: United Nations, p. 146, https://unctad.org/system/files/official-document/aldcafrica2015_en.pdf; UNCTAD (2019) *Economic Development in Africa Report 2019*, New York: United Nations, p. 146, https://unctad.org/system/files/official-document/aldcafrica2019_en.pdf.

10 T. G. Mabilesta, 'South African-Owned Shipping, 2016, thesis, World Maritime University; African Union (2018) '*2050 Africa's Integrated Maritime Strategy Paper*', https://au.int/en/newsevents/20180220/2-0-5-0-africa%E2%80%99s-integrated-maritime-strategy-2050-aim-strategy.

11 PwC (2018) 'Strengthening Africa's Gateways to Trade', www.pwc.com/ng/en/assets/pdf/africas-ports-article.pdf; UNCTAD (2019) *Economic Development in Africa Report 2019*; UNCTAD (2020) *Economic Development in Africa Report 2020.*

12 UNCTAD, *Review of Maritime Trade 2019*, https://unctad.org/system/files/official-document/rmt2019_en.pdf.

13 Equasis, *World Merchant Fleet in 2018*http://www.emsa.europa.eu/equasis-statistics/items.html?cid=95&id=472; UNCTAD (2019) *Economic Development in Africa Report 2019.*

14 UNCTAD, *Review of Maritime Trade 2019*, https://unctad.org/system/files/official-document/rmt2019_en.pdf.

15 Alf Brodin (2009) 'Shipping and the Port Sector in Sub-Sahara Africa', Swedish Maritime Administration/Swedish International Development Cooperation Agency, http://deepseapilotage.se/upload/Listade-dokument/Rapporter_Remisser/EN/2010/ShippingAndPortSectorSub-SaharaAfrica.pdf; EU Investment Bank (2019) 'Blue Economy Finance Principles,' https://ec.europa.eu/maritimeaffairs/sites/maritimeaffairs/files/introducing-sustainable-blue-economy-finance-principles_en.pdf.

16 PwC, 'Strengthening Africa's Gateways to Trade'; UNCTAD (2019); UNCTAD (2020).

17 UNCTAD (2016) *Economic Development in Africa Report 2016*, New York: United Nations, https://unctad.org/system/files/official-document/aldcafrica2016_en.pdf; UNCTAD (2018); UNCTAD (2019).

18 David Noah (2018) 'Documents Required for Cargo Shipment, www.shippingsolutions.com/blog/documents-required-for-international-shipping.

19 The Liberian Registry, www.liscr.com/vessel-registration-department; T. D. Akinyoade and Chibuike Uche (eds) *Entrepreneurship in Africa*, Brill: Leiden, 2017.

20 T. C. Nwokedi (2018) Moses Ntor-Ue Eba, Okonko Iflok and Ndubuisi Leonard, 'Assessment of Shippers…', *LOGI - Scientific Journal on Transport and Logistics* 9

(1), https://content.sciendo.com/view/journals/logi/9/1/article-p70.xml?language=en (accessed 5 April 2020).

21 Tshepo Kgare, Gaël Raballand and Hans W. Ittmann (2011) 'Cargo Dwell Time in Durban': Lessons for Sub-Saharan African Ports', Policy Research Working Paper, No. 5794, Washington, DC: World Bank, https://openknowledge.worldbank.org/handle/10986/3558; Raballand *et al.*, 'Why Does Cargo Spend Weeks in Sub-Saharan African Ports?'; Theophilus Nwokedi, Lazarus I. Okoroji, Ifiok Okonko and Obed Ndikom (2020) 'Estimates of Economic Cost of Congestion Travel Time Delay Between Onne- Seaport and Eleme-Junction Traffic Corridor, November, *LOGI – Scientific Journal on Transport and Logistics* 11(2), doi:10.2478/logi-2020-0013 Project: Blue economy, Shipping, Oil and Gas Logistics Research, www.researchgate.net/publication/346579633_Estimates_of_Economic_Cost_of_Congestion_Travel_Time_Delay_Between_Onne-_Seaport_and_Eleme-Junction_Traffic_Corridor.

22 World Bank Logistics Performance Index; 'Doing Business 2020', https://lpi.worldbank.org/.

23 World Bank, *Doing Business 2020*, 'Comparing Business Regulation in 190 Economies', http://documents1.worldbank.org/curated/en/688761571934946384/pdf/Doing-Business-2020-Comparing-Business-Regulation-in-190-Economies.pdf.

24 CGTN (2019) 'New Depot to Help Ease Congestion in the Port of Mombasa, Kenya', 17 December; *The East African* (2020) 'Mombasa Port Records Increased Cargo Traffic', 6 January, www.theeastafrican.co.ke/tea/business/mombasa-port-records-increased-cargo-traffic-1434234.

25 US CTPAT, 'Minimum Security Criteria Third Party Logistics', US Customs & Border, www.cbp.gov/sites/default/files/assets/documents/2020-Apr/CTPAT%203PLs%20MSC%20March%202020.pdf.

26 Tshepo *et al.*, 'Cargo Dwell Time in Durban'.

27 OECD (n.d.) 'The Competitiveness of Ports in Emerging Markets, The Case of Durban, South Africa', www.itf-oecd.org/sites/default/files/docs/14durban.pdf; 'South African Ports Recognised for Security Excellence', 25 November 2019, www.hellenicshippingnews.com/south-african-ports-recognised-for-security-excellence/.

28 PwC, 'Strengthening Africa's Gateways to Trade'.

29 'Making the Most of Ports in West Africa', Report No. ACS17308, April 2016, www.wacaprogram.org/sites/waca/files/knowdoc/Making%20the%20Most%20of%20Ports%20in%20West%20Africa.pdf; 'Time to Deliver in East Africa, *Port Strategy*, 5 March 2018, www.portstrategy.com/news101/world/africa/potential-and-partnerships.

30 See AfDB (2010) 'Developing Regional Hub Ports', *African Development Report*, Abidjan: AfDB, ch. 6, www.afdb.org/fileadmin/uploads/afdb/Documents/Publications/African%20Development%20Report%202010.pdf; PwC, 'Strengthening Africa's Gateway to Trade'; Victor O. Gekara and Prem Chhetri (2013) 'Upstream Transport Corridor Inefficoencies and the Implications of Port Performance: A Case Analysis of Mombasa Port and Norther Corridor', RMIT University, p. 16, www.researchgate.net/profile/Prem_Chhetri2/publication/263499595_Upstream_transport_corridor_inefficiencies_and_the_implications_for_port_performance_a_case_analysis_of_Mombasa_Port_and_the_Northern_Corridor/links/550796f80cf27e990e07a76e/Upstream-transport-corridor-inefficiencies-and-the-implications-for-port-performance-a-case-analysis-of-Mombasa-Port-and-the-Northern-Corridor.pdf.

31 Jasmina Ovoina (2020) 'Ivory Coast Container Development Enters Next Stage of Development', *Ports and Logistics*, 26 August, www.offshore-energy.biz/ivory-coast-container-terminal-enters-next-stage-of-development/; Par Edialio, 'Budget 2021: Transport 78% of the 57.605 billion Fcfa Reserved for Investments

(Ivory Coast), *Business Directory*, www.africalogisticsmagazine.com/?q=en/content/budget-2021-transport-78-57605-billion-fcfa-reserved-investments-ivory-coast 23 November 2020.

32 UNCTAD (2019).

33 DP World, www.dpworld.com.

34 UNCTAD (2019); UNCTAD 'Review of Maritime Transport.

35 Judd Devermont Amelia Cheatham and Catherine Chiang (2019) 'Assessing the Risks of Chinese Investments in SSA Ports, No 190604, www.csis.org/analysis/assessing-risks-chinese-investments-sub-saharan-african-ports.

36 'Traffic congestion at Apapa', January 2020, www.hellenicshippingnews.com/apapa-gridlock-more-vessels-berth-at-eastern-ports/; Knight Frank, 'Logistics Africa', 2016.

37 Japan International Cooperation Agency (2009) 'The Research on the Cross-Border Transport Infrastructure: Phase 3 Final Report, March, www.icafrica.org/fileadmin/documents/Knowledge/JICA_CrossBorderTransportInfrastructure.pdf.

38 Gylfi Palsson, Alan Harding and Gaël Raballand (2007) 'Port and Maritime Transport Challenges in West and Central Africa', SSATP Working Paper 84, www.ssatp.org/publication/port-and-maritime-transport-challenges-west-and-central-africa.

Part II

Country case studies

7 South Africa

The blue economy experience

Thean Potgieter, PhD

Introduction and setting

The economic and welfare growth the world experienced during the previous century might have been unprecedented, but it was unequal and came at a real cost to the environment. This is to our long-term detriment and concerns about ocean health are glaringly obvious in the African maritime domain. As human beings our existence is inextricably linked to the oceans, which cover more than 70% of the Earth's surface, absorb carbon dioxide, regulate global climate and temperatures, and recycle nutrients. Oceans are crucial to the global economy as they carry more than 80% of global trade, are vital for a food secure future, and provide roughly one-third of the global supply of hydrocarbons, while the coastal regions provide exceptional habitable and leisure spaces.

As governments and business interests increasingly view marine spaces and ocean resources as a means to sustain growth and development trajectories, balancing economic activities with the protection of vulnerable ecosystems must be an issue of increasing importance. The oceans provide 'natural capital' and are 'good business', but as Silver and colleagues warned,[1] due to the unscrupulous and unsustainable exploitation of ocean resources this business potential threatens the environment. The major challenges to our environment, ranging from resource depletion to pollution and climate change, are global in nature and are worsening. Hence, countries and groups of countries must cooperate in balancing preservation and healthy oceans with economic activities to ensure that their decisions and actions support the notion of a truly sustainable global blue economy.

This idea is inherent to the 2030 Agenda for Sustainable Development of the United Nations (UN) as Sustainable Development Goal 14 is aimed at committing member states to 'conserve and sustainably use the oceans, seas and marine resources for sustainable development'.[2] Unfortunately, four years later the UN could not report positively on progress as insufficient environmental care and regulatory frameworks, overfishing and climate change resulted in higher levels of ocean acidification and worsening coastal eutrophication.[3]

These aspects are relevant to South Africa because of its rich ocean resources and the economic growth and development potential associated with its 'oceans economy'. South Africa's coastline – the largest in sub-Saharan Africa – is about 3,900 km long, and by including the Prince Edward Islands its exclusive economic zone (EEZ) measures about 1,553,000 sq km. When the country's continental shelf extension claims are recognized, this could increase to almost four million sq km.[4] However, this vast maritime geography and its inherent wealth is also a scourge as it is difficult to govern and it often provides cover for those engaged in illicit activities. Even though the legal jurisdictions over maritime spaces are stipulated by the UN Convention on the Law of the Sea (UNCLOS) and South Africa's legal framework, they are often ungoverned due to state capacity – the reality is that sovereignty must be exercised in order to be recognized.

In emergent scholarly discourses on the economic use of the oceans, emphasis is placed on diverse issues such as trade and resource wealth, climate change and ocean health, the over-exploitation of dwindling living marine resources, the potential of aquaculture, risks linked to the exploitation of non-living resources, the plight of coastal communities, oceans governance and maritime security threats. However, the link between these diverse aspects and how they fit into a blue economy framework is not always clear. Simply exploiting the ocean for economic purposes is not the hallmark of a blue economy – economic activity must occur in the context of a sustainable marine environment, which is dependent on good governance and maritime security.

The potential contribution that the oceans economy could make to the national development of South Africa grew as a focus area in the literature during the past decade, while linking it to good governance and maritime security received insufficient attention. This topic is often the focus of a separate discourse.[5] As the success of the blue economy is dependent on appropriate maritime governance and maritime security, the interrelated nature of these aspects is a guiding paradigm in this chapter.

South Africa's recent efforts to stimulate the oceans economy occurred under the auspices of the Operation Phakisa Oceans Economy initiative, launched in 2014 ('phakisa' means 'hurry up' in Sesotho). As Operation Phakisa focuses on growth, job creation and economic sustainability, the South African government used the term oceans economy (not blue economy), but the official literature did emphasize 'environmental integrity'.[6] Despite challenges in achieving its ambitious growth projections, Operation Phakisa is of strategic importance and can potentially contribute to the country's socio-economic growth and development.[7] Operation Phakisa did inspire national awareness of the government's efforts to support development and stimulate investments in this sphere of economic activity. However, the focus on maritime issues and oceans governance motivated by Operation Phakisa has since subsided somewhat, which is probably due to the global economic challenges experienced in maritime trade and exploration.

The blue economy in South Africa

Despite the meaning of the term blue economy often being ambiguous or open to interpretation, it is essentially concerned with the economic sectors and policies aimed at ensuring the sustainable use, management and conservation of ocean resources. Of note is a 'measurable' blue economy systems approach that allows standards to be set, while balancing state responsibilities with socio-economic activities and preserving the health of ocean ecosystems.[8] The genesis of the term is probably the June 2012 UN Conference on Sustainable Development held in Rio de Janeiro, Brazil, where the outcome of the blue economy was seen as being similar to that of the green economy – focusing on improved human well-being, social equity, a healthy environment, low carbon and resource efficiency.[9]

Although insufficiently covered in the blue economy literature, the link between development and maritime security is obvious: protecting maritime spaces against illicit activities and threats, good order at sea, and enforcing state authority are crucial prerequisites for regular economic activity. Due to South Africa's dependence on international trade and maritime resources, and the number of state and non-state stakeholders involved, good governance and maritime security are crucial for sustainable growth and development.

The South African government's objective to stimulate growth and development in the maritime sector was clearly articulated in the National Development Plan 2030 (a strategic vision for growth and development published in 2012) and, as introduced above, in the Operation Phakisa Oceans Economy initiative.[10] Operation Phakisa brought together key stakeholders in analysing problems, identifying opportunities, setting priorities and planning. Initially the programme focused on four main areas (maritime transport and manufacturing, offshore oil and gas, aquaculture, and marine protection services and governance), but small harbour development and coastal and marine tourism were soon added.[11]

The initial Operation Phakisa estimations optimistically indicated that the oceans economy could grow from the 2010 level (contributing 54 billion rand to gross domestic product (GDP) and providing 316,000 jobs), to as much as one million employment opportunities and a GDP contribution of up to 177 billion rand by 2033.[12] This would be subject to human resource capacity development, creating an appropriate regulatory framework and developing the potential of local business to compete globally.

Responses to Operation Phakisa were essentially positive as it provided the opportunity to develop the oceans economy for the benefit of all. At the end of 2018 the South African government indicated that the oceans segment of Operation Phakisa (the initiative includes minor projects in other sectors) had secured public and private sector investments of more than 29 billion rand (or US $2 billion), with job creation gains amounting to about 7,093 persons.[13] Investments were mainly in port infrastructure, manufacturing (boat building), aquaculture and offshore oil and gas exploration. The sector as a whole employed about 437,694 people.[14] However, it was soon evident that the initial targets were too optimistic due to a languid domestic economic and investment climate, as well as adverse global conditions with commodity prices placing a damper on investments and predictions for development.[15] Owing to the outbreak of the coronavirus (COVID-19) pandemic the data for 2020 is expected to be much bleaker.

There are divergent views on utilizing ocean resources from around the world and in the South African context. Specific shortcomings of Operation Phakisa include the focus on investments and economic growth, seemingly at the cost of marine conservation.[16] More pessimistic views on the South African blue economy highlight the 'underlying capitalist crisis … associated with over accumulation, globalization and financialisation, as they played out through uneven development, commodity price volatility and excessive extraction of resources'.[17] Ocean health has deteriorated while 'capitalist accumulation', mineral extraction and unsustainable fishing continue; shipping becomes more centralized; tourism more elite-orientated; and more corporate corruption occurs.[18]

Irrespective of ideological, philosophical or economic paradigms, the exploitation and abuse of the oceans for economic gain is a reality that is poignantly illustrated by the activities of capitalist, socialist and other regimes. Globally, from an economic perspective, we need the oceans and their resources to survive. Maritime activities must become more 'blue' and sustainable, and adequate ocean governance and security put in place to ensure that we can reap their bountiful harvests and that future generations are able to enjoy ocean spaces.

Good maritime governance remains essential and constitutes the legislative, institutional and implementation mechanisms aimed at regulating activities in the maritime space and coordinating the involvement of state and non-state actors. As many security threats are non-traditional (such as terrorism, natural disasters, climate change, illegal fishing, marine pollution, maritime safety, smuggling and human trafficking), regulatory frameworks are crucial for sustainable ocean-related economic growth. Frameworks must apply to living and non-living resources, infrastructure, spatial planning, security, environmental management and inter-agency coordination at the national, regional and international level.

As South Africa was relatively slow to strategically prioritize maritime governance, development and security, and has limited capacity to govern its expansive maritime interests, creating appropriate governance frameworks was one of the main focus areas of Operation Phakisa. The Department of Transport was also tasked with coordinating a National Maritime Security Strategy with more than 20 departments and agencies governing the maritime space (including the Department of Environment, Forestry and Fisheries and the South African navy). The strategy had to define roles and jurisdiction; highlight gaps in capacity, budgets and legislation; and plan for greater resources integration. However, this process was interrupted by the COVID-19 pandemic which placed new challenges on governing the maritime domain.

South Africa contributes to continental and regional debates on ocean-related issues because international, regional and bilateral cooperation is regarded as key to a flourishing blue economy. This is specifically relevant for addressing border issues, pollution, overfishing, piracy, access to shipping lanes and ocean governance. South-South cooperation, especially between African states, is useful for promoting blue economic development opportunities and addressing challenges, as insufficient resources and a lack of political will to promote inter-regional frameworks for maritime governance have hindered meaningful development and allowed others to benefit (often unscrupulously) from Africa's rich maritime resources.[19]

The discourse on cooperation and blue economic growth has recently become more important within Africa and the Indian Ocean region, chiefly because the Indian Ocean is moving away from being an 'Ocean of the South to the Ocean of the Centre', to the 'Ocean of the Future'.[20] At a 2018 meeting in Mauritius, the Inter-Governmental Committee of Experts of Southern Africa (under the auspices of the UN Economic Commission for Africa) placed considerable emphasis on collaborative blue economic development, security, exploration, fishing and aquaculture, while also highlighting the importance of political will to create favourable regulatory frameworks and to ensure environmental sustainability.[21]

Utilizing living marine resources

Societies derive much economic value from the exploitation of living marine resources because it is an important source of protein for billions of people around the world. Sadly, though, biologically sustainable fish stocks declined from 90% in 1974 to 65.8% in 2017.[22] In Africa the oceans provide vital nutrition to more than 200 million people, an average of 22% of their daily intake of protein, and an income to more than 10 million people.[23] Greater efficiency in exploiting living resources has caused immense stress to many marine ecosystems and has depleted living marine resources, specifically along the coasts of developing countries with weaker security surveillance and enforcement capacity. South Africa has been no exception to this trend.

Fisheries are important to South Africa, economically and as a source of protein. In 2017 the government estimated the annual worth of the fishing industry at about 8 billion rand.[24] The sector includes small-scale subsistence, recreational and industrial fishing. It is regulated by the Department of Environment, Forestry and Fisheries through a permit system linked to allowable catches.[25] Recorded catches have declined, from a total commercial catch of 674,117 metric ton in 2000, to about 612,200 tons in 2016. The main species are anchovy (260,000 tons), hake (143,000 tons) and pilchard (79,500 tons).[26] The total revenue from ocean fisheries and related services industries was 16.1 billion rand in 2018. Although 68.5% of the total revenue is concentrated in the top 10 companies in the sector, concentration ratios are on a downward trend from about 80% in 2014.[27] As South Africans consume about one-half of the annual national catch (312,000 tons) with a per capita consumption of only about 6.1 kg, the country is a net exporter of fish. Exports of fish were valued at about US $598 million in 2017, exceeding imports of around $424 million.[28]

The fishing industry provides roughly 30,000 direct and 81,000 indirect employment opportunities.[29] Identifying and allocating rights to small-scale fishing communities is a recent positive development as about 29,233 persons in 147 communities are considered to be subsistence fishers in South Africa. They rely mostly on shellfish such as abalone, West Coast rock lobster, mussels, oysters and line-caught fish.[30] As 38% of South Africa's key marine fisheries resources are overexploited and only 29% are optimally exploited, the risks to existing resources are significant and there is no room for industry growth.[31] It should be noted that over-exploited species decreased by 4% between 2016 and 2018, which is ascribed to increased monitoring and assessments by the government.[32]

Due to high profits and growing global markets, illegal, unreported and unregulated (IUU) fishing is one of the most lucrative crimes in the world, considering that annual catches could be up to 26 million metric tons, worth US $23 billion.[33] However, as IUU fishing causes fish stocks to plummet and threatens sustainable ecosystems and biodiversity, it is a major global environmental challenge. Developing countries are specifically at risk: one in four fish sold internationally might be caught illegally in African waters, while up to one-third of the fish sold in the United States could be illegal.[34] Poor governance and capacity constraints are also to blame as African countries often have insufficient control over catches in accordance with licenses provided to foreign vessels.[35]

Despite dwindling global fish stocks, some governments financially support harmful fishing practices and the fishing fleets of many developed countries are subsidized. With their own fishing stocks depleted or heavily regulated, they fish in other regions (such as sub-Saharan Africa) where states have limited law enforcement capacity to prevent IUU fishing. As 2018 data indicates, fuel subsidies among these countries have increased (representing about 22% of fishing subsidies). During the previous decade the People's Republic of China (with the world's largest overseas fishing fleet and the largest amount of subsidies) increased harmful subsidies by 105%.[36] Furthermore, 68% of the world's fishing fleet (roughly 3.1 million vessels) are operated by Asian countries.[37]

In 2016 the Food and Agriculture Organization of the UN (FAO) brokered an accord to stamp out IUU fishing, known as the Agreement on Port State Measures to Prevent, Deter and Eliminate Illegal, Unreported and Unregulated Fishing (PSMA). The 66 parties to the initial agreement included South Africa and 20 other countries from sub-Saharan Africa (see

Annex).[38] By June 2019 FAO indicated that 87 states were signatories to the agreement and that 105 countries were committed to its implementation (an updated country list is not given). Crucial to the agreement is the notion that ports must share information about transgressors, prevent the sale of illegally caught fish, inspect licenses and fishing gear, improve port control to prevent the offloading of fish by suspicious vessels, refuse to resupply and refuel such vessels, and enhance their capacity with technical assistance provided by FAO.[39]

The IUU picture along the South African coast remains stark as some of the once abundant inshore species (such as abalone and rock lobster) have been decimated. Stocks of the Patagonian toothfish in South African waters around the Prince Edward Islands never recovered after being annihilated by IUU fishing in the late 1990s. As South Africa had limited capacity to patrol these rough sub-Antarctic waters at the time, as many as 60 trawlers made lucrative catches of up to 20 metric tons per day (fetching between US $10 and $26 per kg). As it was forbidden to land the fish at South African ports, catches were taken to Mauritius, Mozambique or Namibia.[40]

Trawlers of various nationalities violating catch quotas have been apprehended in South African waters in recent years. During 2016 four Chinese trawlers were apprehended in two incidents. The concerned public expressed outrage at the lack of official action when reports and pictures surfaced of nine identical vessels (with no flags and not appearing on the vessel tracing app) fishing in marine protected areas. All nine trawlers belonged to the same company, and one of them (the *Lu Huang Yuan Yu 186*) was later apprehended. Gear, equipment and cargo were seized, but (conveniently for them) there was no fish on board.[41] Soon afterwards three Chinese vessels with 600 metric tons of chokka (Cape Hope squid) on board that had been caught inside the South African EEZ were impounded in East London. Fines were issued, but they were relatively light (totalling 700,000 rand) and are not an effective deterrent considering the potential value of fully laden trawlers. Infringements continue as the large sea area and incidences of corruption pose severe challenges to law enforcement.[42]

Abalone and West Coast rock lobster are specifically threatened due to their high market value and stocks continue to decline at a catastrophic rate due to high levels of poaching, IUU fishing and poor management of fisheries.[43] The over-exploitation of these species and the involvement of local and international criminal networks are causing the collapse of these stocks. Abalone is a high-value status food item considered a delicacy in Cantonese cuisine[44] and controlled commercial abalone fishery (with a mass quota of about 500 metric tons) was relatively stable for many years. However, due to poaching it was suspended in January 2008 and the current stock is estimated at less than 5% of pristine estimates.[45] Markus Burgener (from the wildlife trade monitoring network, TRAFFIC) estimated in 2018 that syndicates have exported more than 50,000 tons (about 130 million abalones) during the last 25 years, and that the illegal industry has an annual worth of about one billion rand.[46] Reports indicate that smuggling networks involved with illegal commodities such as rhino horn and abalone are operating between South Africa, Namibia and Hong Kong.[47]

In terms of volume West Coast rock lobster comprises only 0.4% of the total South African fisheries catch, but 9.2% in value, due to it also being a valuable export commodity to Asian markets (including Japan, China and Hong Kong). Owing to illegal fishing and poaching natural sources of West Coast rock lobster are in a dire state (currently there exist only 2% of historical levels). The crustacean was placed on the World Wildlife Fund's 'don't-buy red list' in 2016 and it is feared that it could disappear altogether within the next few years.[48] As existing controls do not seem to balance the powerful social, political and economic factors influencing illegal fishing, rock lobster and abalone are close to commercial extinction. Therefore, special efforts are required to safeguard these species and allow them to recover.

As aquaculture provided only 1% of South Africa's fish protein in 2014, it was identified as a potential growth area under Operation Phakisa. Although an unrealistic target of 20,000 metric tons was set for 2019, the government streamlined regulations (specifically for abalone and trout), awarded new leases, initiated investments of close to 500,000 million rand, and supported 36 aquaculture projects.[49] Most of the farms are located in the cold water of the Benguela Current (along the Cape West Coast) and mainly produce abalone, mussels and oysters. South Africa supplies about 21% of the global market for farmed abalone.[50] The aquaculture industry is on an upward curve, but its size is still negligible compared to live fish catches as production amounted to about 5,500 tons in 2016, with about 3,660 tons from saltwater aquaculture.[51] At the same time about 2,500 tons of seaweed were produced, mostly as food for the fast-developing abalone farming industry. Aquaculture still faces many challenges which includes suitable water space, a lack of calm bays, environmental sustainability, security threats, theft and the cost of operating in South African sea conditions. Aquaculture contributes about 4% of the total production in the fishing industry and the potential expansion of the industry could help to remove considerable pressure from the wild fish population.[52]

Fisheries and aquaculture were immediately hard hit by the COVID-19 pandemic. This specifically applied to products dependent on international trade, while demand for fresh fish and shellfish also declined rapidly due to the closure of the food services sector and the loss of tourism. Commercial fishing activities were regarded as essential services in South Africa, but related production, logistics and imports and exports were affected by the lockdown and closure of borders. Although the full impact of COVID-19 remained uncertain at the time of writing, a 'prolonged market downturn is likely'.[53]

Exploiting non-living resources

New technological applications are making ocean exploration and the extraction of offshore oil and gas and seabed mining more economically viable, while considerable new opportunities also exist in renewable energy and desalination. Offshore oil and gas have a mixed legacy in sub-Saharan Africa: despite being a growth area, extraction has rarely brought prosperity to impoverished regions (the 'resource curse'), while considerable security threats and severe environmental degradation are in some cases associated with it. For example, industry sources indicated that in 2019 daily Nigerian losses to theft, spills and in production amounted to 138,400 barrels of crude oil or 7% of total production (roughly US $11.47 million per day) during the preceding 10 years.[54]

When Operation Phakisa commenced, it was estimated that South Africa's fossil fuel reserves could yield nine billion barrels of oil and 11 billion barrels of oil equivalent of gas, which in terms of local consumption represents up to 40 and 375 years of usage, respectively.[55] In early 2019 the French oil firm Total announced that a large new gas condensate discovery in the Brulpadda offshore field in the Outeniqua basin (about 175 km off the coast) was estimated to contain reserves that could yield in excess of one billion barrels of oil equivalent.[56] The Brulpadda discovery sparked much interest and could attract considerable foreign investment as well as generate more than one trillion rand in revenue.[57] The announcement was welcomed as South Africa imports most of its refined petroleum products and crude oil, and as the gas-to-liquid plant in Mossel Bay is running at less than half of its design capacity. Although one-third of the global oil and gas resource is offshore, extracting much of the South African source would pose unique challenges due to the distance from the shore, the depth of the water in some cases as well as the unforgiving sea conditions surrounding the offshore platforms.

Operation Phakisa initially centred on public-private sector cooperation and significant investment promises. Saldanha Bay on the south-western coast of South Africa is one such case. Due to its good infrastructure, deep water and wide expanse, it was identified as a future oil and gas hub and for the repair and maintenance of rigs and offshore support vessels. Port facilities in Saldanha Bay had to be expanded, an offshore supply base had to be created and pipeline routes were identified, but ostensibly due to a more favourable business environment the repair and maintenance activities moved to Walvis Bay in Namibia.

The anticipated growth in the South African oil and gas sector was dealt a serious blow by historically low global prices which by 2016 had halted international production at some marginal wells.[58] Due to the cost of exploration, it was not viable to drill some of the planned offshore wells. This rendered unattainable the ambitious 2019 Operation Phakisa target to add 40,000 jobs and more than 14 billion rand to the economy. According to Masie and Bond, the Phakisa cohort of officials and 'experts' overlooked the broader crises evident in the political economy of oil and shipping which reduced the prospects of major opportunities for South Africa.[59]

The construction of a crude oil storage and blending facility, utilizing the existing Strategic Fuel Fund pipeline and oil jetty continued in Saldanha

Bay. This 13.2-million-barrel facility was not created for the South African market and is used to store and blend different grades of crude oil for international clients. In addition, Africa's largest open-access liquefied petroleum gas import and storage facility used by distributors and bulk consumers was also constructed onshore. However, operations are plagued by a difficult relationship with the Transnet National Ports Authority (TNPA) and financial difficulties.[60]

In early 2020 prospective investors in the South African energy sector indicated interest, given that projects were on a commercially viable scale that good market prices and adequate demand prevailed.[61] However, the imposition of a national lockdown in response to the COVID-19 pandemic shook the energy sector to its core. By 20 April global oversupply and a lack of storage capacity resulted in the benchmark US oil prices sinking to US $0 per barrel and Brent crude oil losing about 70% of its early 2020 value. As the shock reverberated, the economies of the sub-Saharan African oil producing countries, such as Nigeria and Angola, were hard hit.[62] The crisis resulted in oil companies cutting capital expenditure and a greater consolidation of business in the hands of major companies was likely. As a result, investors were expected to consider carefully the associated risks and benefits before investing in South Africa.

The sands along the South African coastline are rich in mineral resources such as cement, aggregates and heavy metals. This amounts to around 30% of the world production in titanium and zirconium. Most of these activities as well as the diamond diving industry (involving the extraction of diamonds from the ocean floor) are based along the South African west coast (and into Namibia). Much of the mining of mineral sands and the prospecting rights granted for marine phosphate are controversial because considerable concern exists about the environmental impact of these mining processes.[63]

Both the demand and opportunities for water desalination specifically in the more arid and drought-stricken western and northern regions of the country are high. There are about 10 small desalination plants, mainly for household use, between Lambert's Bay on the west coast and Richards Bay on the coast of KwaZulu-Natal. Although a number of plans do exist there is no major plant in operation. As one of the shortcomings of desalination plants is their reliance on fossil fuels, an encouraging occurrence is the installation of the first solar-powered desalination plant in sub-Saharan Africa at the small coastal village of Witsand in the Western Cape. It produces about 150 kl of drinkable water per day.[64]

Commerce, trade and shipping

African countries are heavily dependent on maritime trade, but maritime transport is threatened by poor security on the high seas and in ports, as many countries have limited naval, coastguard and air surveillance resources to effectively control their maritime domains. The scourge of Somalia-based piracy was beaten by international and regional cooperation, but the increasing incidence of piracy and kidnapping off the coast of West Africa is of concern.[65]

Despite more than 300 million metric tons of cargo being shipped from South African ports annually and an estimated 12,000 vessels carrying 96% of its exports, South Africa had no locally registered major cargo vessels when the Operation Phakisa oceans economy programme commenced. By 2016 two bulk carriers were registered on the South African Ships Register and by mid-2019 South Africa had five major cargo ships on the local ship register.[66] Local ship registration posed challenges due to high taxation and crew costs, reputational issues and bureaucratic difficulties. However, in order to increase the number of ships registered and to stimulate coastal trade, the South African government is addressing a variety of issues, considering a cabotage policy (for cargo between South African ports to be required to be carried on locally registered ships), and in June 2017 approved the Comprehensive Maritime Transport Policy aimed at facilitating 'growth and development of maritime transport' and international trade in support of socio-economic development.[67]

Operation Phakisa also provided for developing the skills base for shipping, shipbuilding and repair, crews and deck officers. Very optimistic targets were set (18,000 personnel trained between 2014 and 2019), which stimulated existing training centres, resulted in new establishments and programmes, partnerships between the private sector and government, and enhanced training of personnel albeit in smaller numbers. Nevertheless, the main issue will not be the number of trained persons, but rather to ensure continuous placements in industry.

South Africa is not a maritime nation, but it has been poised to become one for some time. What is required is a broader maritime mindset and making good use of South Africa's location, resources and infrastructure to unlock its full potential. Operation Phakisa was an effort to stimulate this through viable partnerships, while studies such as the Council for Scientific and Industrial Research's Road Map for the South African Maritime Sector identified objectives to stimulate a maritime culture, direct research, create enabling governance frameworks, encourage investments, use natural resources sustainably, and enhance safety and security.[68] However, as A. T. Mahan reminded us more than a century ago, geography, economic resources, naval power and government policy are not enough. In order to become a maritime nation sea-mindedness with many of a country's citizens following the way of the sea is also required.[69] Mapping out this road for South Africa undoubtedly poses many challenges during a period of global economic downturn.

Port infrastructure and shipbuilding

The development of ocean-related industrial activities is dependent on good port infrastructure and facilities, such as shipbuilding and repair, maritime services, as well as cargo and passenger terminals. Due to the concentration of high-value articles and commodities many ports in sub-Saharan Africa are prone to crime, trafficking and smuggling; vulnerable to terror attacks, piracy and armed robbery; and contribute to environmental degradation. This highlights the importance of adequate port security and management, while also providing business opportunities for private security.

South Africa has eight major seaports providing easy access to the East, Europe and the Americas. Saldanha Bay (on the Atlantic Ocean) and Richards Bay (the country's most northerly Indian Ocean port) essentially handle South Africa's bulk cargo. Cape Town (at the southern tip of Africa) and Durban (in KwaZulu-Natal) are among sub-Saharan Africa's busiest ports with more than 80 million metric tons of cargo passing through Durban annually. The Eastern Cape boasts three ports: East London, Port Elizabeth and the new Port of Ngqura. Mossel Bay in the Western Cape is the smallest commercial port, but it supports local offshore oil and gas operations. As the poor state of port facilities (specifically dry docks and cranes) has resulted in a decline in shipping repairs and related activities, port upgrades formed part of Operation Phakisa. This specifically entailed improvements to the dry dock and shipbuilding infrastructure, and expansions to port facilities in Durban, Cape Town and Port Elizabeth.[70]

Owing to an international oversupply of ships, 1,400 dry bulk carriers (or 15% of the global fleet) were scrapped in 2016, while larger and more automated ships came into service. These large Capesize ships (bulk and carriers with more than 150,000 DWT capacity) cannot sail through the Panama or Suez canals but must traverse the Cape of Good Hope and Cape Horn.[71] However, due to their shallow berthing capacity the ports in South Africa's four main port cities (Durban, Cape Town, East London and Port Elizabeth) are at a disadvantage, whereas the three deep-water ports that could handle larger ships (Richards Bay, Ngqura and Saldanha Bay) are remote from the major markets while their port and transport infrastructure are insufficient for high container volumes.

The port of Durban is in much need of refurbishment to improve its vessel, container, liquid storage, repair, shipbuilding and vehicle handling capacity. Indeed, this refurbishment is the port's first major infrastructure project in about a century. In response to the increasing container demand at Durban, in 2011 Transnet announced a project to develop a new dig-out port on the location of the former International Airport site (to the south of the current harbour) at an estimated cost of 100 billion rand.[72] The project sparked a profound debate on the political, economic and ecological aspects associated with it. The site was handed over to Transnet at the end of 2012 and operations were supposed to begin by 2018, but the work was deferred to 2026 or even later. Transnet continued with upgrades at the current port of Durban, deepening some of the berths and expanding the container capacity.[73]

The Durban port developments were marred by revelations of extensive corruption.[74] This became evident in 2017 and involved Transnet and the two Chinese suppliers of container-lifting cranes and locomotives. As Masie and Bond indicate, the locomotive deal was funded by a US $5 billion loan that Transnet received in 2013 from the China Development Bank.[75] Transnet overpaid the suppliers of the locomotives by about $1.3 billion, while the infamous network of the Gupta brothers received backhanders of about $400 million on the locomotive deal and about $12 million on the container cranes (a project worth $92 million). Although improvements to the efficiency and competitiveness of Durban harbour is limited by physical constraints (the port is surrounded by the city), future high-cost expansions should be balanced against the port's purpose and anticipated user requirements.[76]

The Coega Industrial Development Zone was established in 1999 at the port of Ngqura. One of its flagships projects is an automotive manufacturing investment valued at 11 billion rand involving the Chinese Beijing Automotive Industry Corporation (BAIC) that aims to produce a sports utility vehicle and a small car. As the project is part of the economic and trade cooperation between China and South Africa, BAIC have a 65% share in the plant and South Africa's Industrial Development Corporation the remaining 35%. The project was behind schedule and, following a further cash injection of 428 million rand, the first unit was delivered from the pilot production line in 2019.[77] Although nothing came of some of the Coega investments, by 2019 they had attracted 45 operational investors, who contributed about 9.53 billion rand in private sector investment, employing 7,815 people, while more than 8,016 jobs were created by infrastructure projects.[78]

Despite South Africa's good port infrastructure in continental terms, the state-owned enterprise managing the ports, Transnet (through the TNPA), is riddled with corruption and inefficiencies. In 2012–13 the total port costs were 360% above global averages and container handling rates were higher than the going rate in developed countries.[79] Port tariffs were dramatically reduced in the following six years and despite container vessel costs being below the global average by 2018, total port costs were still 'relatively high' at about 166% of global averages.[80] As delays and low efficiency levels at busy ports like Durban and Cape Town cause unnecessary cost to shipping companies, they are being addressed.

The refurbishment of small harbours, used for small-scale fishing, aquaculture, diamond mining and leisure activities, was prioritized as part of Operation Phakisa.[81] This includes the redevelopment of 12 fishing harbours in the Western Cape, as well as the ports of Nolloth, Boegoebaai and Hondeklipbaai in the Northern Cape. New small harbours are also planned for Port St Johns in the Eastern Cape as well as Hibberdene and Port Edward in KwaZulu-Natal.[82] Despite the stagnant nature of the South African economy, improved port infrastructure offers opportunities for job creation, growth and development.

During the Second World War, with British ports bombed and the Mediterranean closed to general shipping, all Allied traffic used the Cape Route. The result was that South African ports and shipyards became crucial for naval and commercial ship repairs and maintenance, and for supplying the war effort. Commercial shipbuilding continued after the war and due to the imposition of international sanctions during the apartheid era, naval shipbuilding and repair boomed as various ships were built for the South African navy.[83] However, this industry has since declined.

One of the positive aspects of Operation Phakisa is to stimulate the local shipbuilding industry by making it mandatory for the state to acquire ships from local builders.[84] The Durban-based Southern African Shipyards built 12 tugs for the TNPA and delivered the last of nine new Voith Schneider tugs in 2018.[85] This shipyard, now known by its former name Sandock Austral, is also building a sophisticated new hydrographic survey vessel for the South African navy to replace the legendary SAS *Protea* that has been in service since the early 1970s. The extended impact of the COVID-19 pandemic poses a risk to this project in the form of schedule delays, and a foreseeable funding deficit due to severe cuts in the defence budget.[86]

Other shipyards (such as Nautic in Cape Town) have succeeded in penetrating the sub-Saharan African market by building offshore oil industry patrol vessels for a West African client and have delivered smaller vessels for defence clients in Malawi, Nigeria, Namibia and South Africa.[87] Damen Shipyards Cape Town (a subsidiary of a Dutch company) has constructed more than 40 vessels for the African continent, including two new tugs for the South African navy. In addition, Damen also received a contract for three inshore patrol vessels for the South African navy in December 2017. Construction commenced in February 2019 and the first vessel was expected to be delivered by mid-2021.[88]

Various smaller South African boat builders and industry players have achieved success with producing equipment for naval use, as well as exporting vessels for policing roles and leisure craft. The fact that more than 90% of all the vessels produced in South Africa are exported, together with the manufacturing of new inshore patrol vessels and the highly technical hydrographic survey ship, certainly contributes to the positive reputation of South African shipbuilding. These are tough times and given a post-COVID-19 economic recovery in coastal tourism, shipping and offshore oil and gas, the South African shipbuilding and repair industry can certainly provide many solutions to potential clients.

Nonetheless, South Africa's realistic prospects of becoming a major internationally competitive shipbuilder are at best negligible. This is due to various factors, such as the global oversupply of ships, established and reputable global competition, as well as the limited economy of scale applicable to local shipbuilding. For example, due to limited orders local shipbuilders cannot negotiate the discounted prices for steel and aluminium marine plate that the major players receive.[89] In addition, the South African steel industry has been severely hit by Chinese imports, causing bankruptcy and the shutdown of steel mills. Despite the solidarity rhetoric so akin to BRICS (an informal grouping comprising Brazil, the Russian Federation, India, China and South Africa), emerging market competition has shown unabated ruinous characteristics as far as South African industry, and other interests, are concerned.[90]

Internationally port infrastructure projects remain controversial and some observers claim that South African port developments were white elephant investments as part of a 'blue economy hype', which attracted billions in state subsidies and became subject to corporate corruption.[91] Nevertheless, good port infrastructure can facilitate maritime trade capacity and competitiveness. In the case of South Africa better tariff structures, increased efficiency and improved port infrastructure could stimulate business opportunities, despite the decline in international trade and shipping.

Tourism and recreation

The tourism industry is an important driver of growth and job creation in the South African economy and accounts for about 3% of GDP. The tourism trade balance in comparison with the rest of the world is also positive, and has shifted more in South Africa's favour: in 2011 the surplus was 11 billion rand, which increased to 43 billion rand in 2016, but receded somewhat to 36 billion rand in 2018.[92] Owing to its long and scenic coastline and good infrastructure, South African coastal tourism really took off in the early twentieth century. Although coastal and marine tourism accounts for roughly 40% of overall tourist expenditure and has substantial potential as an earner of foreign exchange, it is dominated by domestic (not international) visitors at a rate of about six to one.[93] Moreover, approximately three-quarters of South Africa's coastal tourism (and its associated spending) is geographically concentrated around two metropolitan areas, Cape Town in the Western Cape and eThekwine (including Durban) in KwaZulu-Natal.[94] Thus, much could be done to enlarge South Africa's international market segment and to allow more South Africans to take a share in coastal tourism.

Efforts to stimulate coastal and marine tourism also formed part of Operation Phakisa. The Coastal and Marine Tourism Implementation Plan (approved by the South African cabinet in 2017) was developed to tackle infrastructure constraints, improve safety and security, support targeted marketing initiatives (such as events, routes and unique attractions), transfer skills, create sustainable coastal developments, and develop small harbours.[95] The blue economy debate also had a positive influence on coastal tourism policies and marine spatial planning as new initiatives are aimed at being environmentally sustainable as well as spreading socio-economic benefits and job creation opportunities more widely.

Regrettably, due to the COVID-19 pandemic in 2020 tourism was 'the worst affected of all major economic sectors' according to the UN.[96] In keeping with global tendencies, South Africa responded by imposing a national lockdown, introducing comprehensive travel restrictions and closed its borders. All domestic and inbound tourism came to an abrupt standstill. Both Cape Town and Durban are popular cruise ships destinations and Cape Town (with its premium passenger terminal) is recognized as a leading international cruise. Construction of an impressive new cruise terminal for Durban commenced late in 2019 and was scheduled for completion in 2021. However, the likely debacles of passengers stranded as a result of outbreaks of COVID-19 on board cruise ships would have eroded the unparalleled growth in leisure cruises and both cities would have been severely affected for a time to come. Preliminary global estimates (at June 2020) were staggering: up to 1.1 billion fewer tourists; losses as high as US $1.2 trillion; up to 120 million jobs at risk; and recovery was expected to be 'slow' at best.[97]

The COVID-19 pandemic resulted in massive revenue and job losses in this growing sector in South Africa. It is estimated that spending on tourism could shrink by at least 171 billion rand, and more than a million tourism-related jobs could be lost.[98] What could be done to mitigate this situation? Government bailouts, an inclusive strategy, and marketing of brand South Africa? Yet the reality is that the COVID-19 pandemic will leave the sector reeling, probably long after the discovery and administration of a vaccine. Tourism was important to the economy and a great job creator. It was one of the first industries to be affected by the pandemic and will probably be the last to recover.

Environmental concerns

Climate change will threaten South Africa's water resources (it is a water-scarce country), food security, health, biodiversity and ecosystems within a socio-economic context that is already confronted by high levels of inequality and poverty. Owing to changing sea temperatures economically important species will migrate, while rising sea levels, more severe storms and coastal development will accelerate coastal erosion and threaten dunes, beaches and coastal habitats.[99]

Much of South Africa's coastline is pristine and rich in biodiversity, but due to adverse sea conditions vessels often come to grief which enhances the threat of pollution. Oil pollution has an adverse impact on the marine environment. It destroys the thermal insulation of seabirds and marine mammals and causes physical smothering (due to the spread of toxic polycyclic aromatic hydrocarbons). South Africa's biggest oil spill occurred in August 1983 when the MT *Castillo de Bellver*, carrying about 250,000 metric tons of light crude oil, caught fire, exploded and broke in half about 122 km north-west of Cape Town. As much of the oil burned or travelled seaward the impact was considered light for a spill of this magnitude, and the most visible impact was that of about 1,500 gannets from a nearby island covered in oil. Of great concern is the fact that the bow section of the ship, that was towed away and sunk about 24 km off the coast, still contains around 100,000 tons of oil.[100]

A much smaller spillage (2,000 metric tons of bunker oil) from the MV *Apollo Sea* that sank off the West Coast in 1994 resulted in the deaths of about 10,000 penguins. When the bulk ore carrier *Treasure* sank in 2000, the 1,300 tons of bunker oil that spilled between Dassen and Robben islands (home to the world's largest African penguin colonies) resulted in 20,251 penguins being covered in oil, of which 2,000 died, and 19,500 had to be relocated.[101] Oil spills require massive clean-up operations and in the longer term chemical toxicity in the ocean and along the coast could cause ecological changes, while habitat loss could lead to the loss of key organisms and species critical to the ecology.

Shipping contributes significantly to marine pollution through the disposal of rubbish, air emissions, sewage, ballast water and oil pollution. The dumping of ballast water and the pumping of bilges have become an important environmental issue as untreated ballast and bilge water can contain harmful fossil fuels and sludge, solvents, detergents and chemicals. Furthermore, alien or invasive marine species can also move from one area to another via ballast water, thereby altering habitats and changing ecosystems.[102] It is estimated that more than two million metric tons of ballast water is dumped off the South African coast annually. Although insufficient research has been done on invasive species in South African waters, the potential to inflict considerable damage is illustrated by the extensive spread of the Mediterranean mussel (*Mytilus galloprovincialis*) that was first noticed in Saldanha Bay during the late 1970s. It soon displaced the native mussel species and changed the natural composition of the rocky shores along the entire West Coast.[103] In line with international best practice South African regulations stipulate that 'ballast water [is] to be exchanged deep sea prior to entering territorial waters'.[104] The South African marine industry has also developed the requisite capacity to install ballast water treatment systems in South Africa.[105] Compliance with and monitoring of these practices remain an issue of global concern.

Conclusion

Although the long-term effects of Operation Phakisa remain to be seen, criticism from a politico-economic perspective highlights conceptual flaws in its planning and execution. Critics note that the initiative was heavily influenced by fast-tracking processes in which ocean resources were essentially seen as commodities in projects aimed at 'capital accumulation', without due consideration of the ongoing crisis in capitalism, oil and shipping.[106] In addition, the emphasis was more on economic gain, while conservation of living resources and ocean health was not prioritized to the same extent.[107]

While fish stocks are under considerable pressure globally, the South African situation is particularly dire. At current levels IUU fishing and poaching will lead to the extinction of certain species in the wild. As the fishing industry is crucial to the South African economy the challenge is to find equilibrium in the use of living resources and the economic benefit derived from it within the context of a sustainable ocean environment. This requires better monitoring of catches and enhanced security aimed at preventing illegal catches in order to allow stocks of threatened species to recover. Aquaculture provides certain growth opportunities, specifically in the nutrient-rich waters of the Benguela Current and has benefited from government support. However, projects must be economically viable and environmentally sustainable.

Other facets of the South African blue economy are also a mixed bag of opportunities, growth and frustrations. Although the oil and gas initiatives were adversely influenced by declining prices and global calls for greener solutions, given higher prices and more lucrative extraction processes, further exploration would become a reality. As infrastructure improvements provide opportunities for enhanced trade, consistent port upgrades are required and port inefficiencies that cause financial losses to operators must be dealt with.

Since South Africa's ocean ecosystems are ecologically threatened, it cannot be 'business as usual' and innovative approaches are required to ensure sustainability. This specifically applies to managing fisheries for recovery, improving marine spatial planning, applying holistic management approaches, and securing more marine protected areas.

Considerations about the South African blue economy would not be complete without acknowledging that the destructive economic impact of the COVID-19 pandemic will doubtless influence the growth of the blue economy. What was difficult to accomplish before COVID-19 will be even more so now. COVID-19 has increased mass vulnerability and could be the bane of sustainable development.

Nevertheless, South Africa has certain attributes that could result in a more positive long-term outlook. Owing to its natural situation, abundant resources, infrastructure and access to international sea lanes, many aspects of the country's blue economy are poised to achieve much better growth and development. However, comprehensive economic growth, greater confidence on the part of investors and good governance are required. It is essential to recognize that all aspects relating to the interaction between human beings, the land and the ocean are interdependent and must be environmentally sustainable.

Annex: parties to the Agreement on Port State Measures

The Agreement on Port State Measures to Prevent, Deter and Eliminate Illegal, Unreported and Unregulated Fishing was approved by the FAO Conference at its 36th session held in Rome, Italy, on 18–23 November 2009 under paragraph 1 of Article XIV of the FAO Constitution, through Resolution No 12/2009 dated 22 November 2009.[108]

Table 7.1 The Agreement entered into force on 5 June 2016, with the following states parties to the Agreement:

1. Albania	24. Guinea	47. Philippines
2. Australia	25. Guyana	48. Republic of Korea
3. Bahamas	26. Iceland	49. Saint Kitts and Nevis
4. Bangladesh	27. Indonesia	50. Saint Vincent and the Grenadines
5. Barbados	28. Japan	51. São Tomé and Príncipe
6. Cabo Verde	29. Kenya	52. Senegal
7. Cambodia	30. Liberia	53. Seychelles
8. Canada	31. Libya	54. Sierra Leone
9. Chile	32. Madagascar	55. Somalia
10. Costa Rica	33. Maldives	56. South Africa
11. Cuba	34. Mauritania	57. Sri Lanka
12. Côte d'Ivoire	35. Mauritius	58. Sudan
13. Denmark (in respect of Greenland and the Faroe Islands – associate member)	36. Montenegro	59. Thailand
14. Djibouti	37. Mozambique	60. Togo
15. Dominica	38. Myanmar	61. Tonga
16. Ecuador	39. Namibia	62. Trinidad and Tobago
17. European Union – member organization	40. New Zealand	63. Turkey
18. Fiji	41. Nicaragua	64. United States of America
19. France	42. Norway	65. Uruguay
20. Gabon	43. Oman	66. Vanuatu
21. Gambia	44. Palau	67. Viet Nam
22. Ghana	45. Panama	
23. Grenada	46. Peru	

Notes

1 Silver *et al.*, 'Blue Economy'.
2 UNECA, *Africa's Blue Economy*, 9.
3 UN, 'Progress of Goal 14'.
4 South Africa, *Green Paper.*
5 Potgieter, 'Oceans Economy', 50–51.
6 Operation Phakisa, 'Unlocking the Economic Potential'.
7 Van Wyk, 'South African Strategic Priority'.
8 Smith-Godfrey, 'Defining the Blue Economy', 63.
9 World Bank, *Potential of the Blue Economy*, 4–5; and UNCTAD, *Oceans Economy*, 2.
10 DPME, 'Operation Phakisa'; and Operation Phakisa, 'Unlocking the Economic Potential'.
11 DEFF, 'Operation Phakisa'.
12 Ibid.
13 DEFF, 'Oceans Economy'.
14 DEFF, 'Sustainable Blue Economy Conference'.
15 Potgieter, 'Oceans Economy', 56, 66–67.
16 Rogerson and Rogerson, 'Emergent Planning', 24; and Potgieter, 'Oceans Economy', 51.
17 Bond, 'Blue Economy', 341.
18 Ibid., 343–44.
19 Masie and Bond, 'Eco-capitalist Crises', 319.
20 Doyle, 'Blue Economy', 1.
21 Republic of Mauritius, 'Southern Africa'.
22 FAO, *State of World Fisheries.*
23 African Union, *Maritime Strategy*, 8.
24 EU-SADC, 'South African Fisheries'.
25 Wepener and Degger, 'South Africa', 110; and FAO, 'South Africa'.
26 FAO, 'South Africa'.
27 Stats SA, *Census of Ocean Fisheries.*
28 Wepener and Degger, 'South Africa', 110; and FAO, 'South Africa'.
29 Ibid.
30 Pillay *et al., Oceans Scorecard.*
31 Wepener and Degger, 'South Africa', 110.
32 Pillay *et al., Oceans Scorecard.*
33 CSIRO, 'Science Tackles Illegal Fishing'.
34 Ibid.
35 Boto *et al.* 'Fighting against Illegal, Unreported and Unregulated (IUU) Fishing', 4.
36 Sumaila *et al.*, 'Updated Estimates'.
37 FAO, *State of World Fisheries.*
38 FAO, 'Parties to the PSMA'.
39 FAO, 'Report of the Second Meeting'.
40 Potgieter, 'Maritime Security', 10.
41 DAFF, 'Inspection of Foreign Fishing Vessel'.
42 Steyn, 'Poaching for Abalone'.
43 WWF, *Ocean Facts*; and Pillay *et al., Oceans Scorecard.*
44 De Greef, 'Poachers and the Treasures'.
45 FAO, 'South Africa'.
46 De Greef, 'Poachers and the Treasures'.
47 Grobler, 'Exposing the Abalone-Rhino'.
48 WWF, *Ocean Facts.*

49 DEFF, 'Oceans Economy'.
50 FAO, 'South Africa'.
51 Ibid.
52 Wepener and Degger, 'South Africa', 110.
53 FAO, 'Q&A: COVID-19'.
54 Reed, 'Nigerian Oil Theft'.
55 Operation Phakisa, 'Unlocking the Economic Potential'.
56 Total Oil Company, 'Significant Discovery'.
57 Persens, 'Expert: Gas Condensate'.
58 Magwaza, 'Commodities Fall'.
59 Masie and Bond, 'Eco-capitalist Crises', 320–21.
60 Tshwane, 'Saldanha Gas'.
61 Smith, 'Investors Still Keen'.
62 Oberholzer, 'Oil Prices Drop'.
63 Wepener and Degger, 'South Africa', 112.
64 Bulbulia, 'Africa's First'.
65 *Africa Times*, 'Global Piracy'.
66 MRA, 'SA Ships Registry'.
67 Department of Transport, *Maritime Transport Policy.*
68 Funke *et al., Research, Innovation*, 16–38.
69 Kennedy, *Rise and Fall*, 1–9.
70 DEFF, 'Oceans Economy'.
71 Kantharia, 'Ultimate Guide'.
72 Samuels, 'Multibillion-Rand Plan'.
73 DefenceWeb, 'No Rush'.
74 Bond, 'Blue Economy', 348.
75 Masie and Bond, 'Eco-capitalist Crises', 316.
76 Bond, 'Blue Economy', 349.
77 Matavire, 'Baic's Coega Plant'.
78 Coega Development Corporation, *Annual Report*, 8.
79 Pieterse *et al.*, 'Supporting Export'.
80 Jacobs, 'Port Tariffs'.
81 Funke *et al., Research, Innovation*, 41.
82 DEFF, 'Oceans Economy'.
83 Potgieter, 'Naval Capacity', 188–92.
84 Bruce, 'Phakisa Floats Hope'.
85 Martin, 'South African Naval Shipbuilding Industry'.
86 DefenceWeb, 'Hotel Sees Progress."
87 Potgieter, 'Oceans Economy', 63.
88 DefenceWeb, 'Project Biro'.
89 Bruce, 'Phakisa Floats Hope'.
90 Masie and Bond, 'Eco-capitalist Crises', 324.
91 Bond, 'Blue Economy', 348.
92 Stats SA, *Tourism*, 17.
93 Rogerson and Rogerson, 'Emergent Planning', 31–32.
94 Rogerson and Rogerson, 'Coastal Tourism', 227.
95 Rogerson and Rogerson, 'Emergent Planning', 33.
96 UNWTO, 'International Tourism'.
97 Ibid.
98 Sithole, 'Seeing Tourism'.
99 Wepener and Degger, 'South Africa', 106.
100 Rantsoabe, 'South Africa's Marine Pollution', 30–31.
101 International Bird Rescue, 'Treasure Spill'.
102 Rantsoabe, 'South Africa's Marine Pollution', 2–10.

103 Robertson, 'Marine Invasions'.
104 SAMSA, 'Marine Notice'.
105 MRA, 'Ballast Water Treatment'.
106 Masie and Bond, 'Eco-capitalist Crises', 320.
107 Potgieter, 'Oceans Economy', 51.
108 FAO, 'Parties to the PSMA'.

Bibliography

Africa Times. 'Global Piracy Down, but Gulf of Guinea Sees a Spike'. *Africa Times,* 15 January 2020. https://africatimes.com/2020/01/15/global-piracy-down-but-gulf-of-guinea-sees-a-spike/.

African Union. *2050 Africa's Integrated Maritime Strategy.* Addis Ababa: Africa Union, 2013.

Bond, Patrick. 'Blue Economy Threats, Contradictions and Resistances Seen from South Africa'. *Journal of Political Ecology* 26, no. 1 (2019): 341–362. https://doi.org/10.2458/v26i1.23504.

Boto, Isolina, Camilla La Peccerella and Silvia Scalco. 'Fighting against Illegal, Unreported and Unregulated (IUU) Fishing: Impacts and Challenges for ACP Countries'. *Rural Development Briefings* 10 (2012). https://brusselsbriefings.files.wordpress.com/2012/10/reader-br-10-iuu-fisheries-eng.pdf.

Bruce, Peter. 'Phakisa Floats Hope for Shipyards'. *Business Day Live,* 25 August 2015. www.bdlive.co.za/economy/2015/08/24/phakisa-floats-hope-for-shipyards.

Bulbulia, Tasmeen. 'Africa's First Solar-Powered Desalination Plant Passes 10,000 kℓ Mark'. *Engineering News,* 27 February 2019. www.engineeringnews.co.za/article/africas-first-solar-powered-desalination-plant-passes-10-000-k-mark-2019-02-27.

Coega Development Corporation. *Integrated Annual Report, 2018–19.* Port Elizabeth: Coega Development Corporation, 2019. www.coega.co.za/DataRepository/Documents/iljJqATjUPhZFnucPTbxNcQ5K.pdf.

Commonwealth Scientific and Industrial Research Organisation (CSIRO). 'Science Tackles Illegal Fishing'. *CSIRO News Release,* 26 June 2017. www.csiro.au/en/News/News-releases/2017/Science-tackles-illegal-fishing?featured=F29EDEB1728C4A92B579C7A5DC28BAD5.

DefenceWeb. 'No Rush to Move AFB Durban as Dig-Out Port Delayed'. *DefenceWeb,* 12 February 2019. www.defenceweb.co.za/featured/no-rush-to-move-afb-durban-as-dig-out-port-delayed/.

DefenceWeb. 'Project Biro on Schedule, with Delivery for Mid-2021'. *DefenceWeb,* 9 June 2020. www.defenceweb.co.za/featured/project-biro-on-schedule-with-delivery-for-mid-2021/.

DefenceWeb. 'Hotel Sees Progress but Funding Is a Concern'. *DefenceWeb,* 11 June 2020. www.defenceweb.co.za/featured/project-hotel-sees-progress-but-funding-is-a-concern/.

De Greef, Kimon. 'The Poachers and the Treasures of the Deep: Diving for Abalone in South Africa'. *The Guardian,* 19 August 2018. www.theguardian.com/environment/2018/aug/19/poachers-abalone-south-africa-seafood-divers.

Doyle, Timothy. 'Blue Economy and the Indian Ocean Rim'. *Journal of the Indian Ocean Region* 14, no. 1 (2018): 1–6. https://doi.org/10.1080/19480881.2018.1421450.

European Union-Southern African Development Community (EU-SADC). '*South African Fisheries'. The Economic Partnership Agreement between the European*

Union and the Southern African Development Community, 2017. https://sadc-epa-outreach.com/images/files/sadc-eu-epa-fisheries-july-2017.pdf.

Food and Agriculture Organisation of the United Nations (FAO). '*Parties to the PSMA', Agreement on Port State Measures (PSMA)*, 2016. www.fao.org/port-state-measures/background/parties-psma/en/.

Food and Agriculture Organisation of the United Nations (FAO). '*The Republic of South Africa', Fishery and Aquaculture Country Profiles*, 2018. www.fao.org/fishery/facp/ZAF/en.

Food and Agriculture Organisation of the United Nations (FAO). *Report of the second meeting of the Parties to the Agreement on Port State Measures to Prevent, Deter and Eliminate Illegal, Unreported and Unregulated Fishing*, 6 June 2019. www.fao.org/3/ca5757en/CA5757EN.pdf.

Food and Agriculture Organisation of the United Nations (FAO). *The State of World Fisheries and Aquaculture 2020: Sustainability in Action*. Rome: FAO, 2020. www.fao.org/state-of-fisheries-aquaculture.

Food and Agriculture Organisation of the United Nations (FAO). *Q&A: COVID-19 Pandemic: Impact on Fisheries and Aquaculture'*, 2020. Accessed 7 June 2020, www.fao.org/2019-ncov/q-and-a/impact-on-fisheries-and-aquaculture/en/.

Funke, Nikki, Marius Claassen, Karen Nortje and Richard Meissner. *A Research, Innovation and Knowledge Management Road Map for the South African Maritime Sector: Charting a Course to Maritime Excellence by 2030*. Pretoria: Council for Scientific and Industrial Research, 2016.

Government of South Africa, Department of Agriculture, Forestry and Fisheries (DAFF). 'Inspection of Foreign Fishing Vessel, Lu Huang Yuan Yu, Registration Number 186'. *Media Release*, 14 May 2016. www.nda.agric.za/docs/media/INSPECTION%20OF%20FOREIGN%20FISHING%20VESSEL%20LU%20HUANG%20YUAN%20YU%20REGISTRATION%20NUMBER%20186..pdf.

Government of South Africa, Department of Environment, Forestry and Fisheries (DEFF). 'South Africa Participates in the Sustainable Blue Economy Conference in Kenya'. *Media Release*, 25 November 2018. www.environment.gov.za/mediarelease/SAparticipatesinthesustainableblueeconomyconference.

Government of South Africa, Department of Environment, Forestry and Fisheries (DEFF). *Oceans Economy Summary Progress Report*, Pretoria: DEFF, Reports, 12 February 2019. www.environment.gov.za/sites/default/files/reports/oceanseconomy_summaryprogressreport2019.pdf.

Government of South Africa, Department of Environment, Forestry and Fisheries (DEFF). *Operation Phakisa – Oceans Economy.* Pretoria: DEFF, Projects and Programmes, n.d. Accessed 11 December 2019, www.environment.gov.za/projectsprogrammes/operationphakisa/oceanseconomy.

Government of South Africa, Department of Planning, Monitoring and Evaluation (DPME). *Operation Phakisa*, Pretoria: DPME, n.d. Accessed 21 September 2019, www.operationphakisa.gov.za/Pages/Home.aspx.

Government of South Africa, Department of Transport. *Comprehensive Maritime Transport Policy for South Africa*. Pretoria: Department of Transport, 2017. www.transport.gov.za/documents/11623/44313/MaritimeTransportPolicyMay2017FINAL.pdf/4fc1b8b8-37d3-4ad0-8862-313a6637104c.

Government of South Africa, Marine Protection Services and Governance. *Unlocking the Economic Potential of South Africa's Oceans: Marine Protection Services and Governance Executive Summary.* Pretoria: Government of South Africa, 15 August

2014. www.operationphakisa.gov.za/operations/oel/pmpg/Marine%20Protection%20and%20Govenance%20Documents/Marine%20Protection%20and%20Govenance/OPOceans%20MPSG%20Executive%20Summary.pdf.

Government of the Republic of Mauritius. 'Southern Africa Intergovernmental Committee of Experts Meets in Mauritius'. *News*, 17 September 2018. www.govmu.org/English/News/Pages/Southern-Africa-Intergovernmental-Committee-of-Experts-meet-in-Mauritius.aspx.

Grobler, John. 'Exposing the Abalone-Rhino Poaching Links '. *Oxpeckers Environmental Journalism*, 2019. Accessed 14 May 2020, https://oxpeckers.org/2019/03/abalone-rhino-poaching-links/.

International Bird Rescue. 'Treasure Spill: South Africa. Helping Save 20,000 Oiled Penguins in Cape Town'. *News and Views*, 30 June 2000. http://blog.bird-rescue.org/index.php/2000/06/2000-treasure-spill-south-africa-2/.

Jacobs, Nicole. 'Port Tariffs Going Down – But Still Relatively High'. *Freight News*, 20 April 2018. www.freightnews.co.za/article/port-tariffs-going-down-still-relatively-high.

Kantharia, Raunek. 2019. 'The Ultimate Guide to Ship Sizes'. *Maritime Insight*, 4 December 2019. www.marineinsight.com/types-of-ships/the-ultimate-guide-to-ship-sizes/.

Kennedy, Paul M. 1986. *The Rise and Fall of British Naval Mastery.* London: Macmillan.

Magwaza, Nompumelelo. 'Commodities Fall Sinks Commercial Vessels Industry', *Business Day*, 26 February 2016. www.businesslive.co.za/amp/bd/economy/2016-02-26-commodities-fall-sinks-commercial-vessels-industry/.

Maritime Review Africa. 'A First for South Africa as Ballast Water Treatment System Installed'. *Maritime Review Africa*, 3 June 2019. http://maritimereview.co.za/article/ArtMID/450/ArticleID/116/A-first-for-South-Africa-as-ballast-water-treatment-system-installed.

Maritime Review Africa. 'SA Ships Registry Welcomes Another Vessel onto Flag'. *Maritime Review Africa*, 20 June 2019. http://maritimereview.co.za/article/ArtMID/450/ArticleID/124/SA-ships-registry-welcomes-another-vessel-onto-flag.

Martin, Guy. 'The South African Naval Shipbuilding Industry'. *DefenceWeb*, 9 February 2018. www.defenceweb.co.za/industry/industry-industry/feature-the-south-african-naval-shipbuilding-industry/.

Masie, Desné and Patrick Bond. 'Eco-capitalist Crises in the "Blue Economy": Operation Phakisa's Small, Slow Failures'. In *The Climate Crisis: South African and Global Democratic Eco-Socialist Alternatives*, edited by Vishwas Satgar, 314–337. Johannesburg: Wits University Press, 2018.

Matavire, Max. 'Baic's Coega Plant Gets R428m Cash Injection'. *City Press*, 9 July 2019. www.news24.com/citypress/Business/baics-coega-plant-gets-r428m-cash-injection-20190704.

Oberholzer, Lizel. 'South Africa: Oil Prices Drop: What Will Fuel the Energy Industry Post-COVID-19?' *Norton Rose Fulbright*, May 2020. www.nortonrosefulbright.com/en/knowledge/publications/9d66757b/south-africa-oil-prices-drop-what-will-fuel-the-energy-industry-post-covid-19.

Persens, Lizell. 'Expert: Gas Condensate Discovery Will Provide Boost for SA Economy'. *Eyewitness News*, 7 February 2019. https://ewn.co.za/2019/02/07/expert-gas-condensate-discovery-will-provide-boost-for-sa-economy.

Pieterse, Duncan, Thomas Farole, Martin Odendaal and Andre Steenkamp. 'Supporting Export Competitiveness Through Port and Rail Network Reforms'. World Bank Policy Research Working Paper 7532, 2016. https://openknowledge.worldbank.org/bitstream/handle/10986/23632/Supporting0exp0tudy0of0South0Africa.pdf;sequence=1.

Pillay, PavsJunaid Francis, Robin Adams, Monica Stassen, Stephanie McGee, Melisha Nagiah, Sindisa Sigam and Kirtanya Lutchminarayan. *2018 Oceans Scorecard: Oceans Facts and Futures.* Cape Town: World Wildlife Fund South Africa, 2018. wwfafrica.awsassets.panda.org/downloads/wwf_sa_ocean_scorecard_2018.pdf.

Potgieter, Thean. 'Maritime Security in the Indian Ocean: Strategic Setting and Features'. *Institute for Security Studies Paper* 236 (August 2012): 1–22.

Potgieter, Thean. 'Building South Africa's Naval Capability: Heyday, Decline and Prospects'. *Journal of the Indian Ocean Region* 10, no. 2 (2014): 183–202. https://doi.org/10.1080/19480881.2014.924218.

Potgieter, Thean. 'Oceans Economy, Blue Economy, and Security: Notes on the South African Potential and Developments'. *Journal of the Indian Ocean Region* 14, no. 1 (2018): 49–70. https://doi.org/10.1080/19480881.2018.1410962.

Rantsoabe, Sibusiso. *Review of South Africa's Marine Pollution Prevention Measures, Particularly those Regarding Vessel-Source Oil Pollution.*" Master of Science Dissertation, World Maritime University, 2014.

Reed, Ed. 'Action Needed on Nigerian Oil Theft'. *Energy Voice*, 8 November 2019. www.energyvoice.com/oilandgas/africa/211750/action-needed-on-nigerian-oil-theft/.

Robertson, Tammy. 'Marine Invasions in South Africa: Patterns and Trends'. *Quest* 11, no. 2 (2015): 44–45. http://academic.sun.ac.za/cib/quest/articles/P44-45.MarineInvasions.pdf.

Rogerson, Christian M. and Jayne M. Rogerson. 'Emergent Planning for South Africa's Blue Economy: Evidence from Coastal and Marine Tourism'. *Urbani Izziv* 30 (2019): 24–36. https://doi.org/10.5379/urbani-izziv-en-2019-30-supplement-002.

Rogerson, Christian M. and Jayne M. Rogerson. 'Coastal Tourism in South Africa: A Geographical Perspective'. In *New Directions in South African Tourism Geographies*, edited by Jayne M. Rogerson and Gustav Visser, 227–247. Cham: Springer, 2020. https://doi.org/10.1007/978-3-030-29377-2_13.

Samuels, Simone. 'Multibillion-Rand Plan for Durban 'Old' Airport'. *IOL News*, 7 November 2011. www.iol.co.za/news/south-africa/kwazulu-natal/multibillion-rand-plan-for-durban-old-airport-1173052.

Silver, Jennifer, Noella Gray, Lisa Campbell, Luke Fairbanks and Rebecca Gruby. 'Blue Economy and Competing Discourses in International Oceans Governance'. *The Journal of Environment & Development* 24, no. 2 (2015): 135–160. https://doi.org/10.1177/1070496515580797.

Sithole, Gugu. 'Opinion: Seeing Tourism beyond Covid-19'. *IOL*, 13 June 2020. www.iol.co.za/travel/south-africa/opinion-seeing-tourism-beyond-covid-19-49302412.

Smith, Carin. 'Investors Still Keen on SA's Energy Sector – As Long as the Project Is Viable'. *Fin24 Economy*, 8 March 2020. www.news24.com/fin24/economy/investors-still-keen-on-sas-energy-sector-as-long-as-the-project-is-viable-20200308.

Smith-Godfrey, S. 'Defining the Blue Economy'. *Maritime Affairs* 12, no. 1 (2016): 58–64.

South Africa Government Gazette. *Green Paper: The South African Policy on Ocean Environmental Management*, Gazette 35783. Pretoria: Government of South Africa, 2012. www.gov.za/sites/default/files/gcis_document/201409/partb.pdf.

South African Maritime Safety Authority. *Marine Notice No. 10*, 9 February 2016.

Statistics South Africa (Stats SA). *Tourism Satellite Account for South Africa, Final 2016 and Provisional 2017 and 2018*. Pretoria: Stats SA, November 2019. www.statssa.gov.za/publications/Report-04-05-07/Report-04-05-072018.pdf.

Statistics South Africa (Stats SA) *Census of Ocean (Marine) Fisheries and Related Services Industry, 2018. Financial and Production Statistics*. Pretoria: Stats SA, 2020. www.statssa.gov.za/?page_id=1856&PPN=13-00-01&SCH=7917.

Steyn, Paul. 'Poaching for Abalone, Africa's "White Gold," Reaches Fever Pitch'. *National Geographic Wildlife Watch*, 14 February 2017. http://news.nationalgeographic.com/2017/02/wildlife-watch-abalone-poaching-south-africa/.

Sumaila, U. Rashid, Naazia Ebrahim, Anna Schuhbauer, Daniel Skerritt and Daniel Pauly. 'Updated Estimates and Analysis of Global Fisheries Subsidies'. *Marine Policy* 109 (November 2019): 1–7. https://doi.org/10.1016/j.marpol.2019.103695.

Total Oil Company. 'Total Makes Significant Discovery and Opens New Petroleum Province Offshore South Africa'. Press release. 7 February 2019. www.total.com/en/media/news/press-releases/total-makes-significant-discovery-and-opens-new-petroleum-province-offshore-south-africa.

Tshwane, Tebogo. 'Saldanha Gas Terminal in Financial Doldrums'. *Mail & Guardian*, 1 April 2019. https://mg.co.za/article/2019-04-01-saldanha-gas-terminal-in-financial-doldrums/.

United Nations (UN). '*Progress of Goal 14 in 2019*'. *Sustainable Development Goals Knowledge Platform, 2019*. New York: UN. https://sustainabledevelopment.un.org/sdg14.

United Nations Conference on Trade and Development (UNCTAD). *The Oceans Economy: Opportunities and Challenges for Small Island Developing States*. New York and Geneva: UNCTAD, 2014. https://unctad.org/system/files/official-document/ditcted2014d5_en.pdf.

United Nations Economic Commission for Africa (UNECA). *Africa's Blue Economy: A Policy Hand Book*. Addis Ababa: UNECA, 2016. www.uneca.org/sites/default/files/PublicationFiles/blueeco-policy-handbook_en.pdf.

United Nations World Tourism Organisation (UNWTO). *International Tourism and COVID-19*, 26 June 2020. www.unwto.org/international-tourism-and-covid-19.

Van Wyk, Jo-Ansie. 'Defining the Blue Economy as a South African Strategic Priority: Toward a Sustainable 10th Province?' *Journal of the Indian Ocean Region* 11, no. 2 (2015): 153–169. https://doi.org/10.1080/19480881.2015.1066555.

Wepener, Victor and Natalie Degger. 'South Africa'. In *World Seas: An Environmental Evaluation*, edited by Charles Sheppard, 101–119. Cambridge: Academic Press, 2019. https://doi.org/10.1016/B978-0-08-100853-9.00006-3.

World Bank. *The Potential of the Blue Economy: Increasing Long-Term Benefits of the Sustainable Use of Marine Resources for Small Island Developing States and Coastal Least Developed Countries*. Washington, DC: World Bank, 2017. https://openknowledge.worldbank.org/bitstream/handle/10986/26843/115545.pdf?sequence=1&isAllowed=y.

World Wildlife Foundation (WWF). *Ocean Facts and Figures: Valuing South Africa's Ocean Economy*. Cape Town: WWF, 2016.

8 From concept to practice

The blue economy in Seychelles

Dominique Benzaken, MSc and
Kelly Hoareau, PhD Cadidate

Introduction

The concept of a sustainable ocean-based or 'blue' economy emerged at the 2012 Rio+20 United Nations (UN) Conference on Sustainable Development,[1] redefining the role of coastal and ocean space in sustainable development, poverty reduction, food security, economic activity and human well-being. It has brought to the fore the threats to ocean health from over-exploitation of marine resources, marine pollution, climate change and maritime security risks such as illegal, unreported and unregulated (IUU) fishing.

While there is no unified definition of the blue economy, at least not in UN terminology, there is general consensus that it is about 'increasing human well-being through the sustainable development of ocean resources, while significantly reducing environmental risks and ecological scarcities'[2]. Scholarly reviews of blue economy policy have revealed four main blue economy narratives, namely oceans as natural capital, oceans as drivers for innovation, oceans as livelihoods and oceans as good business, framing distinct perspectives and conceptual approaches to the implementation of the blue economy in terms of objectives, actors, scale and tools. These narratives have highlighted both synergies, but also potential tension and conflicts.[3] [4] Rather than seeking a standard definition of the blue economy, a context-specific principled approach to the implementation of the blue economy that aims to reconcile economic opportunities, social well-being and the safeguarding of ocean health may be more productive. This requires a paradigm shift in the mindset of those currently exploiting oceans and coastal resources on how ocean and coastal space may be used, how benefits from wealth creation may be shared and ocean health maintained. An integrated blue economy approach can provide a pathway to climate-smart ocean-based sustainable development, consistent with international law and policy.[5] Ultimately, what a blue economy approach brings is a blue lens onto sustainable development, as well as addressing some unique features of the coastal and marine space, and this requires innovative 'blue solutions'.

While the UN Convention of the Law of the Sea provides the overarching legal framework for the use and protection of the ocean, the Sustainable

Development Agenda 2030 and its 17 Sustainable Development Goals (SDGs) provide a useful framework for implementing a blue economy agenda that supports sustainable development. SDG 14 ('life below water') articulates the link between oceans and sustainable development by placing oceans more centrally on the development agenda, thereby creating new opportunities to focus the efforts of the development and finance communities on this essential aspect of sustainable development.[6] Beyond SDG 14, however, the SDGs bring together economic, social and environmental dimensions, which are fundamental to the blue economy, providing a framework and a mechanism for blue economy reporting as part of the implementation of the SDGs. The 2015 Paris Agreement on Climate Change, which acknowledges the role of the ocean in mitigating the impact of climate change, and the detrimental impacts of climate change on marine and coastal ecosystems (e.g. ocean warming, dead zones, ocean acidification) further affirms the need to implement a resilient, sustainable, low-carbon blue economy agenda.

At the African continent level, policy frameworks and practical guidance on the implementation of the blue economy have been instrumental in advocating the role of the ocean in Africa's sustainable development. The African Union (AU)'s Agenda 2063 and the African Charter on Maritime Security and Safety and Development in Africa 2050 (known as the Lomé Charter)[7] both identified the blue economy as a driver towards safe, prosperous and inclusive sustainable development. The UN Economic Commission for Africa's handbook[8] provides policy guidance for implementation at the national level. In 2019 the AU launched its blue economy strategy, which consolidates the role of the ocean in the sustainable development of Africa's future and the implementation of Agenda 2063. The latter envisages an inclusive and sustainable blue economy that will contribute significantly to Africa's transformation and growth.

Since the 2012 Rio+20 conference the blue economy has been embraced and advocated by many Small Island Developing States (SIDS) as a model for the sustainable development of their large ocean domain, and in recognition of their unique dependency on the ocean and vulnerability to environmental and economic risks. A blue economy-centred development model could mitigate some of the challenges confronting SIDS such as small, undiversified economies, high dependency on imports, limited space, skills and capacity, and high unit costs of providing public services.[9] Implementing a blue economy agenda offers the twin potential of realizing economic and social benefits that address these structural challenges, while at the same time improving environmental sustainability and climate resilience. The Republic of Seychelles has played a leading role in promoting a blue economy concept[10] as a central theme for sustainable development since the Third International Conference on Small Island Developing States, held in Apia, Samoa, in September 2014 and has since promoted the blue economy in international and regional forums.[11]

Seychelles at a glance

Seychelles is located in the western Indian Ocean just south of the equator. With an exclusive economic zone (EEZ) of 1.35 million sq km, a land area of just 454 sq km, and comprising 115 islands, Seychelles has a population of around 96,000, mainly concentrated on the three main islands of Mahé, Praslin and La Digue. Seychelles jointly with Mauritius also has jurisdiction over an additional 400,000 sq km of continental shelf (the Joint Management Area) in the Mascarene Plateau region.

Seychelles had a gross national income per capita of US $16,870 in 2019,[12] thanks primarily to the rapid development of its tourism sector, a growth rate of over 3%,[13] less than 4% unemployment (prior to the outbreak of the coronavirus – COVID-19 – pandemic) and a sizable migrant workforce to support its economy.[14] The two main pillars of Seychelles' prosperity are tourism and fisheries. The Seychelles Bureau of Statistics estimated that in 2017 tourism contributed 25.6% to the country's gross domestic product (GDP).[15] Fisheries and seafood processing represented 8% of GDP based only on direct contributions from commercial fisheries, with the Seychelles Indian Ocean Tuna Ltd canning factory alone contributing 6% to GDP, 97% of exports and 8% of employment. Seychelles does not currently commercially exploit oil and gas or minerals, although some offshore exploration is being conducted, and the country is therefore dependent on fuel imports for its energy needs. Seychelles has one of the highest levels of fish consumption per capita (c. 65 kg per annum).[16] With limited agricultural land and an underdeveloped agricultural industry, the country is highly dependent on food imports.[17]

Seychelles' historical development is characterized by a steady increase in income over time, and an economy dominated by a long-standing service sector. Seychelles' Human Development Indicators are broadly in line with national income, with low levels of extreme poverty but substantial income inequality. The Seychellois, however, enjoy comprehensive social services, including free health and education, subsidized housing, utilities, water and waste management services and consistently ranks highest in sub-Saharan Africa on global sustainable development indicators.

Seychelles' unique island and marine environments are characterized by endemism (i.e. species that are unique to the islands) and are of global importance with two UNESCO World Heritage sites, one being the Vallée de Mai on Praslin, and the Aldabra Atoll, famous for its endemic population of giant tortoises and rich marine life. Seychelles has an extensive network of land-based protected areas. In March 2020 13 new areas equivalent to 30% of Seychelles' EEZ were designated as marine protection areas.

Seychelles' approach to implementing the blue economy

Due to its geography and socio-economic characteristics, Seychelles' prosperity directly and indirectly depends on its coastal and marine environment, hence the critical importance of a well-articulated blue economy strategy that can guide national development in a way that respects their ecological integrity. The adoption of a blue economy approach has provided the government with the opportunity to take stock and rethink the development model, starting from the new reality of being a high-income country[18] and looking at the coastal and ocean space as a driver for an integrated approach to ocean-based sustainable development. A blue economy approach casts a strategic long-term 'blue lens' on national development and identifies strategic priorities for action and investment, based on principles of good governance, economic efficiency, sustainability, resilience, innovation and social equity.

Seychelles' blue economy vision is 'to develop a Blue Economy as a means of realizing the nation's development potential through innovation, knowledge-led approach, being mindful of the need to conserve the integrity of the Seychelles marine environment and heritage for present and future generations'.[19] It was first conceptualized under the leadership of former President James Michel, and from 2016 onwards by President Danny Faure. In 2015 the Department of the Blue Economy was established with the mandate to turn the concept into practice and to coordinate the development of a blue economy strategy. The development of the Seychelles Blue Economy Strategic Policy Framework and Roadmap 2018–2030 over the following two-year period benefited from a variety of inputs, including early national consultations and expert studies; a priority-setting exercise facilitated by the Commonwealth Secretariat;[20] a ministerial strategic workshop (held in April 2016); and the ongoing participation of government officials, experts and non-government organizations. The process was supported by a Senior Expert on Ocean Governance funded by the Commonwealth Fund for Technical Cooperation over a two-year period and hosted by the Department of the Blue Economy in the Ministry of Finance and then the Office of the Vice President.

The Blue Economy Strategic Policy Framework and Roadmap: Charting the Future (2018–2030), now known as the Blue Economy Roadmap, adopted by the cabinet in January 2018, provides strategic guidance for implementing Seychelles' blue economy vision across government, private sector and civil society. The Blue Economy Roadmap proposes a prioritized agenda for action and investment up to 2030 to be implemented through four key strategic pillars summarized below:

- Creating sustainable wealth through diversification and progressing the sustainability of existing ocean-based sectors (e.g. fisheries, tourism, ports) focusing on adding value; value chains; sustainability and resilience; taking advantage of Seychelles' comparative advantages and exploring new and emerging sectors (e.g. mariculture, marine renewable energy, biotechnology, digital connectivity, offshore oil and gas); and establishing the policy setting and feasibility of pilot projects.

- Sharing prosperity: ensuring food security and well-being, focusing on improving local production systems and access to markets; reducing dependency on imports and promoting healthy lifestyles; providing access to high-quality education and professional training; creating new jobs and employment opportunities; improving the business environment and encouraging local and international investment, innovation, micro, small and medium-sized enterprises (MSMEs); and a culture of entrepreneurship.
- Securing healthy, resilient and productive oceans through ecosystem service accounting built into economic measures such as GDP; protecting marine and coastal assets, including marine protected areas; implementing blue economy/ocean climate resilience through mitigation (i.e. blue carbon, renewable energy); and adopting adaptation strategies consistent with Seychelles' international obligations.
- Strengthening the enabling environment by establishing blue economy governance and institutional arrangements for cross-sectoral oversight of ocean-based development and protection of ocean environments, and for transparent, inclusive and accountable decision-making and reporting; developing and implementing a Marine Spatial Plan of Seychelles' EEZ; promoting a culture of ocean stewardship and awareness through the use of ocean champions as advocates of sustainable ocean development and protection; developing a research and innovation capability to inform the responsible management of marine and coastal resources and transforming knowledge into development opportunities and productive activities (e.g. biotechnology); financing the blue economy through diversification of Seychelles' funding opportunities, taking advantage of international private sector investors' appetite for investment in sustainability; ensuring greater efficiency of domestic revenue raising mechanisms; incorporating blue economy/ocean risks in national maritime security strategies and regional cooperation to address the impacts of illegal activities (e.g. IUU), resource degradation and improving capacity for Monitoring, Control and Surveillance; strengthening national partnerships between government, industry and civil society and regional partnerships to address issues of common interest; and fostering international advocacy and partnerships to attract technical and financial resources to keep island issues at the forefront of global ocean, climate change, development and finance agendas.

The relationship between the four pillars of Seychelles' blue economy and the SDGs is presented in Figure 8.1.

Progress to date

While it is too early to assess the overall success of Seychelles' blue economy as an integrated framework for ocean-based sustainable development,

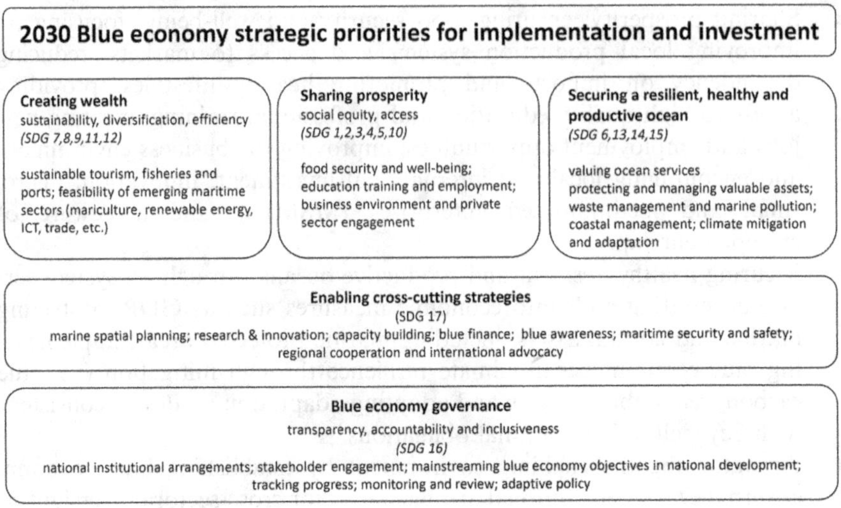

Figure 8.1 Seychelles' strategic framework for the implementation of the blue economy

significant progress has been made through the establishment of an appropriate enabling environment across sectors (see below).

Enabling activities

- The consolidation of institutional arrangements for coordination and stakeholder engagement: A Blue Economy Ministerial Council has been established and is now functional with a broad mandate of strategic oversight, monitoring and review; and a 'Blue Economy Multi-Stakeholder High-Level Forum' to facilitate stakeholder engagement and advice has also been set up.
- The mainstreaming of the blue economy and climate change in the national Vision 2033 and the National Development Strategy 2019–2023.
- The release of a five-year Blue Economy Action Plan in 2020[21] to advance priority actions incorporated in the Blue Economy Roadmap, including communication, strengthening coordination, establishing a blue economy satellite account and progressing strategic projects on research and development, carbon neutrality, marine litter and maritime security and regional integration.
- In 2020 30% of Seychelles' EEZ was designated as a marine protection area.[22]
- Seychelles' first sovereign blue bond worth US $15 million was issued in 2018 with partial guarantees from the World Bank and the Global Environment Facility Trust Fund (GEF) to finance the transition to sustainable fisheries.[23] The Development Bank of Seychelles is disbursing

$12 million in the form of loans to Seychellois for eligible activities and the Seychelles Conservation and Climate Change Adaptation Trust (SeyCCAT)[24] is disbursing $3 million in grants for fisheries governance projects.

- The SeyCCAT's Blue Grant Fund has had four competitive calls for proposals relating to stewardship of Seychelles' ocean resources and marine conservation.[25]
- Policy reform of doing business is underway to improve private sector engagement in the blue economy. The National Institute of Science, Technology and Innovation (NISTI) is preparing a Business, Technology and Innovation incubator strategy to support local entrepreneurs, and in 2020 an African Development Bank grant (linked to this) was funding MSME capacity development and knowledge relating to marine biotechnology).
- The James Michel Blue Economy Research Institute (BERI), hosted at the University of Seychelles (UniSey), has successfully offered two online blue economy courses, in partnership with the Commonwealth of Learning, and UniSey has launched a new MSc in Marine Science and Sustainability. A special issue on the blue economy was published in 2019 by the *Seychelles Research Journal*. BERI has also contributed to the High Level Panel for a Sustainable Ocean Economy. Local marine and climate research has also been undertaken with local and international partners, including Nekton, the One Ocean Hub, the Food and Agricultural Organization of the UN's Nansen Programme, the Western Indian Ocean Marine Science Association, the Ministry of Environment, Energy and Climate Change, the Green Islands Foundation (GIF), the Marine Conservation Society Seychelles, and the Fishermen and Boat Owners Association.
- The SIDS Youth AIMS Hub-Seychelles, Wise Oceans, and the Department for the Blue Economy have initiated blue economy youth programmes. Some of these initiatives have been funded by the SeyCCAT Blue Grant Fund.

At the sectoral level

- The EU Seychelles Fisheries Agreement was signed in 2020.[26] As part of the Fisheries Comprehensive Plan, electronic monitoring will be compulsory for all industrial fishing vessels within Seychelles' EEZ, with electronic monitoring currently being piloted. A joint venture investment, Central Common Cold Store Ltd, a central common platform for fish storing, processing and logistics efficiency is being built.
- The Seychelles Mariculture Master Plan, policy and regulations are being finalized, and infrastructure development and trials have progressed.

- The Seychelles Port Victoria redevelopment project to upgrade port facilities including an environmental management plan is about to start (pending the COVID-19 pandemic).
- The Seychelles Tourism Master Plan has been updated to include a stronger sustainable tourism component (Destination 2023),[27] and partnerships have developed between the private sector and civil society in, for example, marine education and coral restoration.
- A National Climate Change Policy: A Sustainable, Climate Resilient and Low-Carbon Seychelles was approved by the cabinet in May 2020, and the development of a 4-MW floating solar photovoltaic (PV) farm is under way, along with other plans for solar PV projects. A climate finance advisor funded by the Commonwealth Secretariat has been recruited.
- A Seychelles Coastal Management Plan 2019–2024[28] has been developed and more work is being done to linked large-scale restoration to support hotels and other infrastructure impacted by coastal erosion. There are also internationally funded coral reef projects being led by local organizations such as Nature Seychelles[29] and the Seychelles National Parks Authority.[30]
- A ban on plastic has been in place since 2018 and this was followed by a ban on other single-use items. Community-based groups have been active in beach cleaning and awareness raising of marine pollution and the impact of litter on the marine environment, including a successful clean-up of the Aldabra Atoll (a World Heritage site).[31]

At the regional and international level

- The GEF is funding the UN Development Programme-Joint Management Area Standalone Demonstration Project addressing ocean governance, data management and marine spatial planning for the development of a management framework for the Seychelles-Mauritius extended continental shelf in the Mascarene Plateau region.[32].
- Seychelles was nominated blue economy champion for the AU and is the Commonwealth Blue Charter champion for marine protection areas.
- As a founding member of the Fisheries Transparency Initiative (FiTI),[33] Seychelles is hosting the FiTI global headquarters and is preparing its first report.

The planned mid-term review of the implementation of the Blue Economy Roadmap in 2024 will be an opportunity to formally evaluate implementation of the blue economy as an integrated approach, address challenges that have been met along the way and adapt to emerging issues and changing circumstances. The coronavirus (COVID-19) pandemic is expected to have made a significant impact, affecting investment priorities and the rate of progress overall. This is discussed further below.

From concept to practice: points for reflection

Although the concept of the blue economy has been debated extensively in international forums, examples of the implementation thereof are few. Reflection on how a sustainable blue economy as an integrated approach to ocean-based sustainable development is implemented can provide valuable insights. The experience of Seychelles provides such an opportunity. Below are some initial observations on the key building blocks to consider that could inform a global dialogue and a blue economy community of practice.

Leadership and political will

Strong leadership and political will are central to the effective implementation of a blue economy agenda. Seychelles' advances in implementing important aspects of a blue economy have achieved global visibility through the early leadership, vision and advocacy of former President Michel, and currently President Faure as a blue economy champion for the AU. It has led to important partnerships and support from international organizations and countries, bringing expertise and resources to the government of Seychelles in support of developing its blue economy and marine conservation agenda. However, consensus across the political spectrum on the pertinence of a blue economy approach to sustainable development has been slow and divergences on implementation remain.

Generating stakeholder support

As noted above, Seychelles has looked to the ocean and coastal spaces to meet its needs for generations, as have most island nations. Implementing a blue economy agenda is a long-term project. The endorsement of the Blue Economy Roadmap by the cabinet in 2018 marked the start of integrated thinking about the place of the ocean in the country's development and provided a strategic framework and prioritized agenda for implementation. While progress has been made in establishing an appropriate enabling environment across sectors, implementation will take time and resources as well as continued momentum and commitment across government and stakeholders. Progress in generating awareness and support for the blue economy has been slow; however, a positive change in mindset is emerging. The renewed focus of the Government Blue Economy 2020 Action Plan on communication and awareness strategies is a welcome development. Developing consistent messages targeting the range of blue economy stakeholders will help to address the communication gap and improve stakeholder confidence and constructive engagement. A number of civil society-related initiatives have played a significant role in raising awareness of the ocean and the importance of its protection for the future.

Understanding comparative advantages

The foundation of Seychelles' blue economy is its unique coastal and marine environment. It is the basis of today's tourism and fisheries industries and livelihoods, and of future development opportunities. As a small island state that is vulnerable to global economic and climate change shocks, the sustainable use and protection of its coastal and ocean environments are essential to its resilience and as well as its identity as an island. Seychelles' comparative advantages are based on its unique geography, and its socio-economic and natural assets. The Seychelles 'Blue Economy Brand' focuses is articulated through innovation and diversification of its economy, high-value marine products, high-quality tourism experiences and services and, most importantly, a strong focus on sustainability and conservation management, which in turn acts as a premium attraction for visitors and investors alike, and positions Seychelles as a key actor in the global blue economy. The Seychelles 'Blue Economy Brand' is the thread running through the Blue Economy Roadmap. Maintaining ocean health and sustainability credentials is not without challenges, hence the critical role of the Ocean Council and the high-level multi-stakeholder forum to ensure good communication and a coherent approach across government, private sector and civil society.

Fostering an integrated blue economy agenda

Institutional coordination and policy coherence

As an integrated framework, the Seychelles Blue Economy Roadmap required institutional arrangements to align government institutions and policy tools for coordinated implementation. These are the mandate of the Ocean Council which provides strategic oversight and policy coherence, facilitates cross-sectorial blue initiatives, ensures value addition to sector-based implementation, mobilizes resources, and tracks progress on implementation of the Roadmap. The High Level Multi Stakeholder Forum aims to provide high-level engagement in the implementation of the blue economy. The location of the Department for the Blue Economy as the primary coordination mechanism in central government agencies (first the Ministry of Finance and now the Office of the Vice President) plays an essential part in legitimizing an integrated approach and cross-sectoral collaboration. The Blue Economy Roadmap's strategic priorities for action and investment, developed with cabinet ministers and through consultations with ministries, are designed to guide, harmonize and inform sector-based policies and planning. Streamlining the policy and institutional landscape can pose a challenge to the status quo and sector-based model of policy implementation. Overcoming such a challenge requires demonstrating how the blue economy adds value and addresses areas of common interest or concern.

The marine spatial planning process as a tool to facilitate cross-sectoral decision-making

The Seychelles Marine Spatial Plan, developed through a multi-stakeholder process, allocates zones and sets the rules applicable to ocean-based development and the protection of Seychelles' coastal and marine ecosystems. An ocean authority is under consideration to implement the Plan and to advance integrated ocean management, ocean governance and a sustainable blue economy. Implementation of the Plan requires ongoing resources, some of which are provided by the proceeds of the Seychelles debt swap and disbursed competitively by the independent Seychelles Conservation and Climate Adaptation Trust. However, given the size of Seychelles' EEZ, strategies to mobilize additional resources will be needed (see below).

Monitoring and reporting

Although there is a centralized reporting framework in place for the implementation of the SDGs and some sector-based monitoring and reporting mechanisms, a comprehensive monitoring and reporting framework for the blue economy has yet to be developed and implemented. It is as much an issue of digital communication technology and services (e-government) as it is an issue of strategic leadership and coordination across knowledge-based institutions and government reporting processes. Strengthening the national statistical capacity of government through a dedicated blue economy node for data collection and analysis and developing a user-friendly interface would facilitate the implementation of the blue economy, assist in reporting on results and financial flows, and secure new resources. Ultimately such systems contribute to the transparency and accountability of financial flows and are required by donors. They also support adaptive management by providing the basis for regular review and updating of policy implementation as emerging issues and innovation arise.

Resources for implementation

Most developing nations experience no difficulty in attracting funding for policy development. Attracting long-term support for implementation is not so straightforward. It requires transparency and accountability, as well as financial and political stability. It also requires evidence of institutionalization of policy outcomes within government structures, processes and resources, a challenge for any SIDS with limited human capacity, trained personnel, expertise and financial resources.

Following the global financial crisis of 2008, Seychelles embarked on macro-economic and financial management reforms, which resulted in significant growth and a sound revenue base (mainly generated from tourism and telecommunications). As a result, Seychelles graduated as a high-income country in 2015.[34] Achieving a stable economic and political environment has contributed to the success of Seychelles in attracting investment in support of its marine conservation and blue economy agenda, as evidenced by the Seychelles debt swap in 2015 and the Seychelles blue bond in 2018. However, these successful and high-profile initiatives that were developed with the support of international partners are not alone sufficient to fully implement the blue economy agenda. The Blue Economy Roadmap identified the importance of an integrated investment strategy targeting strategic priorities and considering strategies to building domestic revenue, attracting public and private investments and a suite of financial instruments, which has yet to be fully implemented. Such implementation would identify investment gaps, drive policy reforms to maximize domestic revenue and donor finance and facilitate private sector investment. It would also provide a range of financial mechanisms for resourcing implementation up until 2030, when reporting on SDG 14 will be due. Establishing a dedicated blue economy satellite account, as suggested in the Blue Economy Action Plan, will be essential in tracking financial flows linked to the blue economy and demonstrates its contribution to Seychelles' GDP.

Building an inclusive sustainable blue economy agenda

Private sector engagement

Engaging the private sector effectively depends on the business environment and the professionalization of a local business sector. Diversifying traditional sectors and exploring new and emerging opportunities (e.g. mariculture) offer local business opportunities. Developing the capacity and entrepreneurship of MSMEs and addressing the barriers to doing business, such as access to credit and land, would lead to job creation and encourage innovation. The proceeds of the blue bond disbursed as subsidized loans for blue business ventures have yet to be fully taken up, making the case for professionalizing a local business sector through a combination of business training, incubators to attract emerging entrepreneurs and fostering professional associations. A more diversified blue economy focused on value added and ancillary services would reduce the pressure on ocean resources and improve economic resilience. Recent initiatives focusing on business reform and entrepreneurship are steps in the right direction.

A qualified workforce

As a small country with an extensive EEZ, Seychelles is faced with major implementation challenges both in terms of human resource capacity, trained personnel and expertise. Seychelles' labour force is small and lacks many of the skills needed to take advantage of current and future blue economy opportunities arising from economic diversification, and as a result it is dependent on international specialist skills, in particular for managerial and specialized posts. At the same time, Seychelles is dependent on a foreign labour for unskilled occupations, such as construction or fish processing. Despite efforts that to date include the establishment of professional training centres and the University of Seychelles, the need for a plan to improve Science, Technology, Engineering and Math (STEM) education, along with technical and vocational training remains a challenge. This situation combined by (perceived) limited career opportunities leads to a brain drain that the country cannot afford. Education initiatives that support developing the blue economy and retain skilled labour would allow access to blue economy opportunities that ensure ecological sustainability and social equity.

Developing a research and development capability

Thanks to its unique coastal and marine environment, Seychelles has attracted international researchers for many years, but the ability of Seychelles to build its own research and development (R&D) capability is limited. Much of the research to date is the result of multi-stakeholder and donor-funded (usually with local co-financing) initiatives, which have supported national sustainable development-related priorities. Furthermore, there is currently no national research strategy (nor a formal budget allocated to research) that prioritizes, manages and incentivizes local research development and uptake. The Blue Economy Research Institute (BERI), the National Institute for Science, Technology and Innovation, the Ministry of Environment, Energy and Climate Change, the Department for the Blue Economy and local partners are advocating for a more cohesive research strategy and agenda. The limited R&D capacity, the lack of a prioritized research agenda and the scarcity of comprehensive and up-to-date data hinder adaptive management and implementation of a sustainable and inclusive blue economy. Strengthening partnerships between international, regional and local R&D organizations provides a mechanism to build such capability, to share innovative learning and to mentor local early career researchers. BERI has provided a focal point for such partnerships, fostering local ocean-related research and the development of successful educational tools. Similarly, the SeyCCAT, as an independent trust, is supporting the blue economy, marine conservation and climate change projects. This is attracting international donor interest in capitalizing the SeyCCAT, in particular philanthropic organizations who are interested in co-designing and financing activities related to the marine conservation, climate change, sustainable livelihoods and the Marine Spatial Plan (e.g. blue carbon projects).

Regional cooperation and international advocacy

Many of Seychelles' blue economy opportunities and challenges are transboundary and require regional cooperation for their implementation, from the management of tuna fisheries, the imposition of maritime security to the pooling of R&D capacities and the provision of training opportunities. Regional governance in the Indian Ocean region is complex, despite several blue economy initiatives such as the Nairobi Convention, the Indian Ocean Commission and the Indian Ocean Rim Association. The AU's 2019 Blue Economy Strategy may assist in federating islands and neighbouring countries around key sustainable development and blue economy issues, building on regional institutions and existing regional programmes. The treaty concerning the joint management of the continental shelf in the Mascarene Plateau region between the Government of the Republic of Seychelles and the Government of the Republic of Mauritius that came into force in 2012 is a pertinent example of such bilateral cooperation.

Seychelles' advancement in implementing its blue economy is not without challenges. As a small island nation, it remains vulnerable to external economic shocks, climate change, natural disasters and illegal activities, and more recently the COVID-19 pandemic. Marine pollution, overfishing, IUU, piracy, global warming and ocean acidification pose risks to Seychelles' marine and coastal ecosystems and resources, upon which its economy and well-being depend. International advocacy in global UN forums and regional cooperation at the Indian Ocean Tuna Commission, the Indian Ocean Commission and the AU, for example, can help to mitigate some of those risks. However, mainstreaming resilience into all aspects of the blue economy remains an ongoing challenge. Building strong partnerships around areas of common interest will strengthen the negotiating power of small island nations to generate greater benefits from their ocean resources while ensuring the long-term protection of their coastal and ocean assets.

Impact of the COVID-19 pandemic

The impact of the COVID-19 pandemic on Seychelles' blue economy has yet to be assessed, but it is expected to be significant as a result of the temporary shutdown of international travel, changing investment priorities, delays in implementing planned activities and the loss of resource capacity. Seychelles' credit rating was downgraded by Fitch Ratings from BB in 2019 to B+ in 2020 in the wake of the shutdown. The economy was predicted to contract by an estimated 14% in 2020 owing to the decline in tourism.[35] Thanks to a loan from the International Monetary Fund and other donors for budget support,[36] Seychelles is expected to maintain its debt at a manageable level and to be able to repay its creditors. Seychelles has issued solidarity bonds to help to cover the deficit caused by the pandemic to address its impact on the health and economic sectors. The COVID-19 pandemic has highlighted Seychelles' vulnerability to external shocks, its dependence on international tourism, and the importance of diversifying its revenue base. The pandemic has also brought several national systemic issues to the fore as well as providing an opportunity for collective and innovative action. A full impact assessment of the COVID-19 pandemic on the blue economy as part of a national effort will be invaluable.

Conclusion

The blue economy as a concept has attracted global attention since 2012 as a sustainable development pathway for small islands and nations with large ocean domains. However, examples of successful implementation are few.

A blue economy agenda as a pathway to sustainable development is a long-term endeavour, which is influenced by a nation's aspirations and its economic, social and environmental circumstances. The key point here is that one size does not fit all. The desirability and pace of implementation are dependent on many factors, both internal and external. However, the principles of sustainability, resilience and inclusiveness, which Seychelles' blue economy aspires to, ought to underpin any sustainable blue economy strategy.

Seychelles' experience of implementing an integrated blue economy agenda emphasizes the value in establishing an enabling environment that allows a sustainable and inclusive blue economy to evolve. It includes the government's institutional arrangements, stakeholder engagement, the principles of an agreed national vision, a stable political and economic environment that is attractive to investors, and a willingness to innovate. It is an ambitious agenda for a small island nation and this chapter illustrates some of the challenges it faces. These challenges are a function of its scale, unique characteristics and overall vulnerability.

Financial, technical and human resources will remain a key challenge to successful implementation of a blue economy. Nonetheless, Seychelles continues to contribute to the global conversation on blue economy, advancing its own brand of a sustainable blue economy, thanks in part to its successful international advocacy and partnerships. The COVID-19 pandemic will be a setback for the country, as it will for many SIDS, and the recovery process, including the contribution of its sustainable blue economy, will be a test of its resilience.

While this chapter has identified some of the building blocks for the successful implementation of a blue economy, progress presented is not comprehensive nor evaluated, as a formal review would be, but is indicative of the type of actions and challenges encountered which could inform a blue economy global dialogue and a community of practice.

Acknowledgments

The lead author would like to thank the Commonwealth Fund for Technical Cooperation which funded her position as Senior Ocean Governance Expert during her work on developing Seychelles' Blue Economy Strategic Policy and Roadmap, and for the constructive input and comments from government officials.

Notes

1 United Nations Conference on Sustainable Development, Rio+20. https://sustainabledevelopment.un.org/rio20.html
2 Organisation for Economic Co-operation and Development (OECD). 2016. *The Ocean Economy in 2030*. Paris: OECD. https://doi.org/10.1787/9789264251724-en (accessed 30 July 2020).
3 Jennifer J. Silver, Noella J. Gray, Lisa M. Campbell, Luke W. Fairbanks and Rebecca L. Gruby. 'Blue Economy and Competing Discourses in International Oceans Governance'. *The Journal of Environment & Development* 24, no. 2 (2015): 135–60. www.jstor.org/stable/26477597 (accessed 30 July 2020).
4 Michelle Voyer, Genevieve Quirk, Alistair McIlgorm and Kamal Azmi. 2018. 'Shades of Blue: What Do Competing Interpretations of the Blue Economy Mean for Oceans Governance?' *Journal of Environmental Policy & Planning* 20 (5): 595–616. https://doi.org/10.1080/1523908X.2018.1473153.
5 For example, Meg R. Keen, Anne-Maree Schwarz and Lysa Wini-Simeon. 2018. 'Towards Defining the Blue Economy: Practical Lessons from Pacific Ocean Governance'. *Marine Policy* 88 (February): 333–41. https://doi.org/10.1016/j.marpol.2017.03.002.
6 UN Conference on Oceans (5–9 June 2017), 'Our Ocean Our Future Call for Action'.
7 African Union. 2014. 2050 Africa Integrated Maritime (AIM) Strategy. https://au.int/sites/default/files/treaties/37286-treaty-0060_-_lome_charter_e.pdf (accessed 30 July 2020).
8 United Nations Economic Commission for Africa (UNECA). 2016. Africa's Blue Economy: A Policy Handbook. Addis Ababa: UNECA. http://repository.uneca.org/handle/10855/23014.
9 The Commonwealth, the Government of Seychelles. Seychelles' Blue Economy. Strategic Policy Framework and Roadmap: Charting the Future (2018 2020). https://beri.unisey.ac.sc/wp-content/uploads/Seychelles%E2%80%99-Blue-Economy-Roadmap.pdf (accessed 30 July 2020).
10 M. Agrippine. 2014. *The Blue Economy: Seychelles' Vision for a Blue Horizon*. Ministry of Foreign Affairs. https://books.google.sc/books?id=cSMDogEACAAJ.
11 For example, Seychelles hosted two summits on the margins of the Abu Dhabi Sustainability Week (in 2014 and 2016) in partnership with UNESCO-IOC, IRENA, the Green Growth Initiative, and the Commonwealth.
12 https://data.worldbank.org/country/SC and www.worldbank.org/en/country/seychelles/overview (accessed 30 July 2020).
13 Speech by the Minister of Finance. 2018. Victoria: Government of the Republic of Seychelles, Ministry of Finance.
14 Seychelles Bureau of Statistics (2017). www.nbs.gov.sc/news/91-seychelles-in-figures-2017.
15 Tourism Master Plan, Part 2, 2023. www.tourism.gov.sc/lib/TOURISM_MASTER_PLAN_PART_2_TOURISM_SECTOR_STRATEGY_DESTINATION_2023.pdf.
16 Government of the Republic of Seychelles. National Food and Security Policy (2013).
17 National Bureau of Statistics Report, 2017.
18 Seychelles graduated as a high-income country in 2015 and thus became ineligible for ODA funding, despite its vulnerability as an island.
19 Government of the Republic of Seychelles (2018). Seychelles Blue Economy Strategic Policy Framework and Roadmap Charting the Future (2018–2030).
20 Commonwealth Secretariat. (2016). Alternative Future Visions for the Seychelles Blue Economy. London: Commonwealth Secretariat.

21 Government of the Republic of Seychelles and UNECA. 2020. Blue Economy Action Plan.

22 Both actions were conditions of the Seychelles debt swap with the Club of Paris, facilitated by the Nature Conservancy. www.seychellesnewsagency.com/articles/2463/Swapping+Seychelles+debt+for+ocean+conservation++milestone+agreement+reached+with+Paris+Club+creditors (accessed 31 July 2020).

23 The proceeds of the blue bond combined with the World Bank SWIOFish project are an example of blended public and private finance towards the transition to sustainable fisheries.

24 SeyCCAT was established to administer funds derived from the Seychelles debt swap. www.seychellesnewsagency.com/articles/9976/Seychelles+launches++million+blue+bond+to+support+marine+projects (accessed 31 July 2020).

25 For details of projects completed, see https://seyccat.org/projects/ (accessed 31 July 2020).

26 www.seychellesnewsagency.com/articles/11850/Seychelles%2C+EU+agree+on+new+fishing+deal+worth++million+euros+for+island+nation (accessed 30 July 2020).

27 Government of the Republic of Seychelles. 2020.Tourism Master Plan, Part 2 (Destination 2023). www.tourism.gov.sc/lib/TOURISM_MASTER_PLAN_PART_2_TOURISM_SECTOR_STRATEGY_DESTINATION_2023.pdf

28 www.gfdrr.org/sites/default/files/publication/seychelles-coastal-management-plan.pdf.

29 http://natureseychelles.org/what-we-do/coral-reef-restoration (accessed 30 July 2020).

30 http://natureseychelles.org/knowledge-centre/news-and-stories/738-press-release-nature-seychelles-launches-coral-reef-restoration-toolkit-developed-in-the-seychelles (accessed 30 July 2020).

31 www.seychellesnewsagency.com/articles/10786/%2C+flip-flops+among+the++tonnes+of+trash+cleaned+up+on+Seychelles+Aldabra+Atoll (accessed 30 July 2020).

32 Under the treaty concerning the joint management of the continental shelf in the Mascarene Plateau region between the Government of the Republic of Seychelles and the Government of the Republic of Mauritius (the Joint Management Area, JMA).

33 See https://fisheriestransparency.org (accessed 31 July 2020).

34 African Economic Outlook. 2017. www.africaneconomicoutlook.org/seychelles/ (accessed 31 July 2020).

35 www.seychellesnewsagency.com/articles/12875/Seychelles%27+credit+rating+downgraded+due+to+tourism+shutdown (accessed 31 July 2020).

36 www.seychellesnewsagency.com/articles/12879/Int%27l+Monetary+Fund+gives+Seychelles++million+in+emergency+assistance (accessed 31 July 2020).

9 Managing the blue economy

A case study of Tanzania

Francis Mwaijande, PhD

Introduction and methodology

This chapter examines Tanzania's perspectives on managing the blue economy in response to the African Union (AU)'s goal to exploit Africa's marine resources highlighting in particular the challenges and opportunities to increase youth and women's participation in the blue economy within the framework of the African Union's blue economy agenda encapsulated in its 2050 Africa's Maritime Strategy to enhance a sustainable blue economy that can improve Africans' well-being while significantly reducing marine environmental risks as well as ecological and biodiversity deficiencies. This is also aligned to the United Nations Sustainable Development Goal 4, the aim of which is to 'conserve and sustainably use the oceans, seas, and marine resources for sustainable development'.

Tanzania's second Five-Year Development Plan (2015/16–2020/21) notes the sectoral policy of the Ministry of Livestock and Fisheries which focuses on developing aquatic and marine resources and utilizing the oceans for socio-economic development and improving the livelihoods of the country's inhabitants. Tanzania is fortunate to border the Indian Ocean, which is a resource for developing and benefiting from the emerging blue economy.

In the African context, the blue economy operates according to the principles of inclusiveness, environmental sustainability, innovation, governance and dynamic business models.[1] Geographically, Tanzania lies on the east coast of Africa and has a coastline of about 1,424 km in length stretching from Tanga to Mtwara and flanking the Indian Ocean, which is rich in natural resources, beaches, harbours and ports. Tanzania's coastal areas could be some of the major drivers for the socio-economic transformation of the country and human development.[2]

This chapter discusses Tanzania's management and governance of the blue economy in the Indian Ocean and focuses on the fisheries, marine infrastructure, tourism, energy, industry and trade sectors. Data for this study were collected using multi-research methods comprising reviews of Indian Ocean Rim Association (IORA) documents, national policy documents and legislation that together formed an understanding of the blue economy in

Tanzania. This material was triangulated with interviews conducted with officials from the Ministry of Livestock and Fisheries, the Ministry of Infrastructure Development, the Ministry of Natural Resources and Tourism, the Ministry of Energy and the Ministry of Industry, Trade and Investment, who provided valuable information about Tanzania's economic activities relevant to the blue economy. The data (both quantitative and qualitative) were synthesized to present the current state of the blue economy in the selected sectors in Tanzania using SWOC (strengths, weaknesses, opportunities and challenges) analysis to evaluate and understand Tanzania's blue economy.

The fisheries sector

Tanzania's fisheries sector falls under the remit of the Ministry of Livestock and Fisheries, and can be divided into the following subsectors: marine and inland capture fisheries, aquaculture (marine and freshwater) and fish processing. Marine fish capture is carried out in the country's offshore exclusive economic zone (EEZ) and in the deep waters of the Indian Ocean. Marine fishing activities contribute only 1.71% of national gross domestic product (GDP), provide employment directly and indirectly for four million people, and are a source of foreign exchange. In 2018 Tanzania earned US $239,680,014 from exports of fish and fish products.

Marine fisheries

Tanzania's marine fisheries can be broadly divided into three main types; artisanal (small-scale coastal) fisheries, prawn fisheries and offshore fisheries which cater for local consumption and the export market. Marine fishing activities are conducted within Tanzania's territorial waters, which extend up to 12 nautical miles in the EEZ, and up to 200 nautical miles from the Indian Ocean shoreline. Fishing takes place in water depths of less than 500 m and within 40 nautical miles of the Indian Ocean coastline in an area of 64,000 sq km. Almost all of Tanzania's coastal communities are engaged in fishing activities.

Artisanal fisheries

The marine artisanal fleet operates mostly with small dug-out canoes between 3 m and 5 m in length, and wooden planked boats ranging from 6 m to 15 m in length. Smaller vessels are powered by paddle and sail, while larger vessels are powered by advanced inboard and outboard motors. According to interviews conducted with ministry officials, in 2019 there were 7,664 vessels involved in fishing activities which land their catch at 257 sites along the Tanzanian coast.[3] It should be noted that the number of vessels has been increasing at an average rate of 280 vessels per year. As much as 95% of the marine fish production comes from the territorial sea. The marine fisheries production appears to have remained relatively stable, ranging between 43,000 and 55,000 metric tons per annum.[4] Artisanal fisheries supply a modest trade in a number of higher value species such as crabs, lobsters, octopus, shrimps and squid which are sold to the local market.

Offshore fisheries

Only a few Tanzanian vessels are capable of exploiting the marine resources found in the EEZ despite the potential for development for blue economy in this area because the country does not own commercial vessels. Since 1998 the government of Tanzania has licensed foreign-flagged fishing vessels to operate in the EEZ. About 74 vessels were licensed in 2014 from four countries: Spain (14 vessels), France (two vessels), Seychelles (seven vessels) and the Republic of Korea (two vessels).[5] It is in the Indian Ocean waters that fish stocks of skipjack, yellowfin and bigeye tunas, and other large pelagic fish such as sharks, swordfish and marlins are captured. Table 9.1 shows fish species characterized by their migratory pattern, and Tanzania is one of several countries in whose EEZ such fish stocks are found but which remain underexploited. The best fishing season for large pelagic fish in Tanzanian waters is during the north-east monsoons between November and March.

Cases of illegal, unreported and unregulated fishing in Tanzania

With the support of the independent, African-based not-for-profit organization Stop Illegal Fishing, the government of Tanzania is strengthening its capacity to halt illegal, unreported and unregulated (IUU) fishing activities in the EEZ. In partnership with the international conservation organization Sea Shepherd, Tanzania has increased its ability to monitor IUU fishing. According to data from Stop Illegal Fishing, there have been numerous reports of IUU fishing in Tanzanian waters, as the shown by the following examples.

Case study 1: In 2018 a Chinese-flagged fishing vessel, the *Tai Hong No 1*, was discovered carrying a cargo of shark fins that far exceeded the 50 carcasses on board, thus violating Tanzanian law that the number of fins must correspond to the number of trunks.[6]

Case study 2: In 2018 a Malaysian-flagged fishing vessel, the *Buah Naga No 1*, was caught violating fisheries regulations. Evidence of shark finning was found, in contravention of Tanzanian regulations and Indian Ocean Tuna Commission conservation and management measures.[7]

Case study 3: In 2009 Tanzania collaborated with South Africa, Mozambique and Kenya in the monitoring of marine resources in the western Indian Ocean 180 nautical miles off the coast of Tanzania. The South African environmental protection vessel, *Sarah Baartman*, spotted a tuna longliner, *Tawariq 1*, that had no flag. The *Tawariq 1* switched off its radar and accelerated away, ignoring calls to stop for an inspection. After several attempts

Table 9.1 Types of fish found in Tanzanian waters and estimated volume of annual catch

Species (scientific name)	*Local name (Swahili)*	*Common name (English)*	*Catch (kg)*	*No. of fish*	*No. of hauls*
Encrasicholina heteroloba	Dagaa mchele	Shorthead anchovy	1,397,02	499,078	2
Trichiurus lepturus	Mtepa/Antepa/ Mkonge	Largehead hairtail	697,1	1,227	4
Upeneus taeniopterus	Mkundaji	Fin-stripe goatfish	620,3	24,584	6
Decapterus macrosoma	Ngulangula/Msumari	Shortfin scad	559,4	18,840	8
Saurida undosquamis	Bumbura	Brushtooth lizardfish	374,08	3,036	25
Rexea prometheoides	–	–	353,37	4,583	8
Secutor insidator	Kotwe/palawe/kifuu	Pugnose ponyfish	307,58	19,823	3
Carangoides malabaricus	Kolekole	Malabar trevally	278.95	6,178	8
Argentina euchus	–	–	231,72	4,025	8
Leiognathus elongatus	Kotwe/palawe/kifuu	Slender ponyfish	192,99	16,418	7
Myctophidae	–	Lanternfishes	173,76	29088	8
Nettastoma parviceps	–	Duck-billed eel	165,59	1,957	2
Polymixia berndti	–	Beardfish	153,42	6,497	11
Centrophorus granulosus	–	Gulper shark	136,42	67	3
Zenion sp.	–	Zeniontid fish	124,4	1,232	1
Aluterus monoceros	–	Unicorn leatherjacket	115,41	46	2
Gazza minuta	–		105,32	5570	3
Decapterus russelli	Ngulangula/Msumari	Indian scad	103,31	1,462	8
Polysteganus coeruleopunctatus	–	Seabream	102,53	97	4
Abalistes stellatus	Kikande/tundu/vidui	Starry triggerfish	93,09	148	9
Himantura jenkin-sii/wrong name	–	–	89,86	8	2
Correct name Dasyatis jenkisii	Nyenga/Taa	Pointed-nose stingray			

Species (scientific name)	*Local name (Swahili)*	*Common name (English)*	*Catch (kg)*	*No. of fish*	*No. of hauls*
Himantura uarnak correct name	Nyenga	Leopard stingray			
Scomberomurus commerson	Nguru-Maskati	Narrow-barred Spanish mackerel	85,44	54	4
Leiognathus berbis	Kotwe/palawe/kifuu	Berber ponyfish	84,48	41,814	5

Source: Author's elaboration of Johannessen *et al.* (2018), with author's additional Swahili translations.

to intercept, the *Sarah Baartman* eventually caught up with *Tawariq* 1 which was carrying more than 260 metric tons of fresh and frozen fish including tuna and shark fins with no valid licence to fish in Tanzania's EEZ.[8]

Policy implications

Action is needed to halt the threat of IUU fishing in Tanzanian waters due to insufficient surveillance activities. The following are policy implications for the government of Tanzania.

- Strengthen monitoring/patrol units within the existing partnerships between the government of Tanzania and other IORA member states to protect marine resources from the threat of IUU fishing.
- Strengthen partnerships with international bodies that can transfer technology and experience to Tanzania to support Tanzania's initiatives against IUU fishing.
- The Tanzanian government should build its own vessels and should encourage the private sector to build vessels that will follow sustainable marine resources requirements.
- The government of Tanzania has introduced new legislation in the form of the Deep Sea Fisheries Management and Development Act, 2020, to address a range of challenges facing the deep-sea fishing industry. One of the major challenges includes IUU fishing. Christensen argues that IUU fishing is a global phenomenon that has devastating environmental and socio-economic consequences for developing countries such as Tanzania.[9] With a monitoring capacity in the EEZ of about 241,453 sq km, Tanzania has not been able to curb IUU fishing. The provisions of the new legislation will improve monitoring, control and surveillance in the EEZ. The law also prescribes charges against unauthorized fishing vessels. In its 2020 election manifesto the Chama Cha Mapinduzi (Revolutionary Party of Tanzania) pledged to purchase five commercial deep-sea fishing vessels to promote the blue economy by 2025.

Fish processing and marketing

Fish storage and processing capacity

Developing and managing the blue economy requires a sophisticated infrastructure including commercial fishing vessels, fishing ports, storage facilities and processing facilities. At the time of writing this chapter, there were no dedicated port facilities for industrial fishing vessels in Dar es Salaam harbour. However, one area of the port (Berth 6) has been earmarked for use by the fisheries sector by the Ministry of Livestock and Fisheries. The quay has the length and depth to accommodate large ships and offers potential for the landing of fish, and for transshipment to refrigerated transport vessels. There is also potential for constructing cold storage and fish processing facilities in the Indian Ocean ports. This is a promising economic development area for Tanzania's blue economy.

Upstream markets

Upstream markets at landing sites and fish market levels are managed by local governments (city, municipal and district councils) with some collaboration with the Ministry of Livestock and Fisheries. Fish processing industries are mainly privately owned and are regulated by the Ministry of Livestock and Fisheries. Other upstream production systems include privately registered vessels engaged in deep-sea fishing, aquaculture and mariculture.

Downstream markets

Downstream markets include the marketing system and distribution channels for marine fish caught off Tanzania's coast. The markets are managed by the central government in collaboration with the countries of destination for the fishery products through bilateral agreements which set standards between the respective countries and Tanzania. The downstream markets comply with standards set by the European Union when exporting fishery products to EU member countries. Data from the Ministry of Livestock and Fisheries show that the marketing of fishery products is controlled by the private sector whereby small fish traders at various markets along the Indian Ocean coastline obtain their livelihoods. However, organized fish exporting is done by licensed exporters.

SWOC analysis of the fisheries sector

Strengths in undertaking blue economy activities in the fisheries sector is when properly implemented activities are improving countries' economies by earning more foreign exchange and providing more employment. The fact that Tanzania possesses marine waters covering a large area measuring 61,500 sq km which is about 6.4 % of the country total land area as well as an EEZ measuring a total of 223,000 sq km gives it strength in developing economically important blue economy activities.

Weaknesses in undertaking blue economic activities

Tanzania lacks the technological capabilities of utilizing fisheries resources in the EEZ which has deep waters. Fishing in this zone requires large vessels with sophisticated technologies which the country does not possess; as stated above Tanzania uses foreign licensed fishing vessels. Currently, Tanzania's fishing fleet is superseded by foreign fishing fleets in the EEZ. A lack of trained personnel with the necessary skills to operate fishing vessels in the EEZ also hinders Tanzania in undertaking blue economy activities effectively in the deep sea. A lack of capital is another major limiting factor.

Opportunities for undertaking blue economy activities

Tanzania's fishing activities in the EEZ could help to develop the blue economy, but require better management, governance and maritime security. There is also a need to establish fish processing and storage facilities along the Indian Ocean coastline.

There are also promising opportunities for seaweed farming and processing in Zanzibar which so far have not been fully exploited. The majority of the unprocessed seaweed is exported. The potential for growing eucheuma seaweed in Zanzibar is being explored because it is indigenous to Zanzibar's waters.[10] It has been observed that the island is typhoon-free and possesses a large reef acreage facing the Indian Ocean, and it is this that makes the island favourable for seaweed farming. A proposal was submitted to the Zanzibar government, which welcomed the idea of seaweed farming. This is one of the unexploited potential activities for the blue economy in Zanzibar.

Challenges for undertaking blue economic activities in the fisheries sector

Climate change, pollution and overfishing are the main threats which affect undertaking blue economic activities in the fisheries sector because they threaten the sustainability of fisheries resources and aquatic environments.

The energy sector

According to the Ministry of Energy, one of the key blue economy growth areas is oil and gas exploration. Oil and gas exploration in Tanzania has been underway since the 1950s, with the first natural gas discovery being made in early 1974 when the government increased its investment in exploration and this resulted in the discovery of 2.5 trillion cu ft of natural

Table 9.2 SWOC analysis of Tanzania's fisheries sector

Strengths	*Weaknesses*
Availability of local, regional and international markets Presence of national laws, regulations and policies that encourage the blue economy through partnerships between the government and private sector to develop facilities and infrastructure Government collaboration with development partners in fisheries Investment in fish processing technologies through the Tanzania Fishing Corporation Enabling environment: policy and legislation	Insufficient essential infrastructure (fishing ports, docks) Poor private sector investments Lack of robust fish stock assessments and other related studies Insufficient capacity to enforce monitoring, control and surveillance activities Inadequate high-technology fishing vessels and fishing gear
Opportunities	*Challenges*
Availability of an EEZ measuring 223,000 sq km Expansion of fish catch, processing and exports Increase in population results in demand for fish protein Fishing trade, fish processing industries East African landlocked countries market Fisheries and aquaculture technological improvement	Illegal fishing Lack of appropriate technologies and infrastructure for fishing, fish handling, processing and distribution Regional and international competitive policies for the private sector including issues related to royalties, levies and taxes High investment costs in EEZ Environmental pollution

Source: Author's compilation.

gas at Songo Songo Island in southern Tanzania.[11] It was expected that natural gas would make a large contribution to electric power generation in Tanzania. By 2017 natural gas contributed about 625.5 MW of the total power installed capacity (1,450 MW), followed by hydropower (609 MW) and liquid fuel (188.5 MW).[12] It is also expected to be utilized in manufacturing fertilizer for the development of the agriculture sector. However, the full potential of the natural gas subsector that could benefit all Tanzanians remains underdeveloped.

In 1982 a second discovery of about 5 trillion cu ft of natural gas was made at Mnazi Bay in the Mtwara region. The commercialization of these discoveries triggered onshore and offshore exploration and there followed a string of significant onshore discoveries at Kiliwani, Mkuranga and Ntorya.[13] Moreover, a substantial discovery of natural gas was made in 2010 in a deep offshore block in the Indian Ocean. The deep offshore discoveries compelled Tanzania to formulate policy, legal and regulatory frameworks to guide the development and proper management of its natural gas resources. These included the National Energy Policy (2013), the Oil and Gas Revenue

Management Policy (2015), the Petroleum Act (2015) and the Oil and Gas Revenue Management Act (2015). However, from an economic perspective Tanzania's oil and gas potential remains underexploited.[14]

Upstream market in the energy sector

The upstream activities in the energy sector are largely undertaken by the Tanzania Petroleum Development Corporation (TPDC), which is a designated institution for the National Oil Company and the various multinational petroleum companies that have been engaged in exploration activities during the past 60 years. Oil exploration has resulted in the abundant gas discoveries at Songo Songo, Mnazi Bay (see above) and in the southern deep-water basin in Ruvuma region. The cumulative seismic coverage in the public domain is approximately 100,000 km; 70,000 km offshore and 30,000 km. Between 2002–07 exploration licenses were to offered to various multinational petroleum companies including Petrobras (Block 5, 2004), Ophir Energy (Block 1, 2005), Ophir Energy (Blocks 3 and 4, 2006), Statoil (Block 2, 2007), Dominion (Block 7, 2007), Petrobras (Block 8, 2012). These resulted in 70,000 live km of 2D seismic data and 15,000 sq km of 3D seismic data. Further exploration followed the licensing and drilling of wells by the BG Group (Blocks 1, 2 and 3), Statoil (Block 2) and Petrobras (Block 5), which led to significant discoveries of gas in Blocks 1, 2 and 3. The upstream market in the energy sector is also referred to as the production system in the petroleum and gas sector. The organized production of petroleum is effected through the bulk procurement of oil through the Petroleum Bulk Procurement Agency. Registered vessels ship the oil to the Tanzania Ports Authority.

Downstream market in the energy sector

Downstream activities include distributing 423 MW of natural gas annually to the Tanzania Electric Supply Company, industries such as Twiga Cement, and more recently, in June 2020, for private domestic use. The downstream market in the oil and gas sector refers also to the marketing system and distribution channels of petroleum and gas in Tanzania. The downstream markets are managed by the executive agencies of the government including the Petroleum Bulk Procurement Agency, the Energy and Water Utilities Regulatory Authority, oil marketing companies and liquefied petroleum gas (LPG) companies which are engaged in distribution within the country and to its landlocked neighbours. The following include (but are not limited to) companies engaged in the petroleum and gas value chain including the Tanzania Petroleum Development Corporation, Gas Supply Company (GASCO), oil marketing companies, LPG companies, Oil Com and Lake Oil.

The contribution of the energy sector to the national blue economy

The natural gas produced at Songo Songo and Mnazi Bay is a source of clean energy for both industrial and domestic use whereby 48 industries and over 400 households are connected to the natural gas infrastructure. Natural gas contributes over 55% of the country's total electricity mix (892.7 MW out of 1602.3 MW). Table 9.3 indicates the connected industries and their contribution to industrialization in Tanzania.

Since commercialization of natural gas began in Tanzania in 2004, Tanzania has made savings of over US $13.1 billion by using natural gas instead of fuels such as heavy fuel oil for generating electricity. Other benefits include the increased employment of local citizens in the oil and gas value chain activities. For example, blue economy activities have created jobs for Tanzanians including local service providers and suppliers of goods and services in the oil and gas sector.

Table 9.3 The contribution of natural gas discoveries to industrialization

Industry	*Location*
Goodwill Ceramic Mkuranga	Coastal region
Dangote cement factory.	Mtwara region
Coca-Cola factory	Dar es Salaam region
Twiga Cement	Dar es Salaam region

Source: Author's interviews with Ministry of Energy officials.

Table 9.4 SWOC analysis of the energy sector

Strengths	*Weaknesses*
Indian Ocean Potential for huge discoveries of natural gas, existence of the Petroleum Bulk Procurement Agency	Under-exploration Technical know-how, skills Human resource capacity
Opportunities	*Challenges*
Exploration blocks (open acreage) Business opportunities in the mid- and downstream segment of the petroleum value chain	Oil price fluctuations Outbreak of coronavirus (COVID-19) pandemic

Source: Author's compilation.

The tourism sector

Tourism is one of the key drivers of growth in the blue economy and is managed by the Ministry of Natural Resources and Tourism. The nexus of tourism to the blue economy is that tourists choose to visit countries which provide an infrastructure that is conducive to tourism. Some of Tanzania's main attractions are its clean and sustainably managed beaches and well-built hotels. This type of infrastructure has a bearing on the number of tourists entering the country, and these in turn contribute to generating foreign exchange. The upstream market in the tourism sector is organized and promoted through the Tanzania Tourism Board and targets the high-end market in Europe, the United States, the People's Republic of China and the Russian Federation. Such visitors arrive by air and stay in tourist hotels located in Tanzania's coastal areas. Table 9.5 shows the number of tourist arrivals in Tanzania between 2005 and 2018.

The largest market share of tourists travelling to Tanzania include the European Union countries (61.5%), the United States (14.1%), India (7.3%) and China (5.0%). Following the outbreak of the coronavirus (COVID-19)

Table 9.5 Number of tourists visiting Tanzania

Year	*Number of tourist arrivals*	*Annual % change*	*Receipts (US$ million)*	*Receipts (Tanzanian shillings, millions)*
2005	612,754	4.8	823.05	929,058.85
2006	644,124	5.12	950	1,079,137.01
2007	719,031	11.62	1,198.76	1,290,542.26
2008	770,376	7.14	1,288.70	1,520,429.11
2009	714,367	–7.27	1,159.82	1,511,704.59
2010	782,699	9.5	1,254.50	1,767,967.85
2011	867,994	10.89	1,353.29	2,107,613.85
2012	1,077,058	24	1,712.75	2,691,929.18
2013	1,095,884	1.7	1,853.28	2,962,653.40
2014	1,140,156	4	2,006.32	3,316,647.59
2015	1,137,182	–0.26	1,901.95	3,774,443.94
2016	1,284,279	12.1	2,131.57	4,640,641.05
2017	1,327,143	3.2	2,258.96	5,040,191.55
2018	1,505,702	13.5	2,595.59	5,855,154.48

Source: Author's interview with Ministry of Tourism and Natural Resources officials.

pandemic, the market share is likely to be unpredictable in the near future due to air travel restrictions imposed in visitors' countries of origin.

The main blue economy related tourist activities include ecotourism, coastal tourism and water sports tourism. Coastal and ocean-related tourism comes in many forms such as dive tourism, maritime archaeology, surfing, cruises, ecotourism and recreational fishing. The tourism industry offers unique tourist products along coastal beaches including maritime archaeology, surfing, cruises, ecotourism, and recreational fishing. Both the public and private sectors are engaged in marketing the blue economy.

While the government provides an enabling environment including the formulation of a National Tourism Policy[15] for developing tourism, which allows private sector share on tourism infrastructure and business. Tanzania has made considerable strides towards sustainable coastal and marine governance through developing and implementing various policies and strategies including the Integrated Coastal Environment Management Strategy in Zanzibar and mainland Tanzania. Promoting marine planning tools such as marine spatial planning as a keystone to attaining blue economy in the country whereby training on marine spatial planning have been conducted to various stakeholders.

Tanzania's tourism industry has been negatively affected by the COVID-19 pandemic that has paralysed the global travel industry, and this has caused fluctuations in the blue economy too. According to statistics published by the Bank of Tanzania, for the quarter ending September 2020 the number of tourist arrivals had declined from 174,057 in the same quarter in 2019 to 12,867, with the majority (61.5%) coming from various European countries followed by the United States (14.3%).

The government of Tanzania implemented a number of policy actions in response to the pandemic, with the goal of mitigating the effects of COVID-19 on society and the economy. Measures included closing public and private schools, colleges and universities, suspending all public meetings, sporting events and international passenger flights. The Ministry of Health, Community Development, Gender, Elderly and Children provided public health education to combat COVID-19 including hand washing with disinfectant and running water, using hand sanitizers and wearing face masks in public areas along with the use of traditional therapy practices. By the beginning of the third quarter of 2020, the government of Tanzania claimed that the number of cases of COVID-19 was declining nationally. This prompted the reopening of flights that are vital for bringing tourists into the country.

The government remained cautious about the disease, and therefore it advised that travellers, residents and non-residents alike, entering or leaving the country had to undergo testing for COVID-19. The outcome has been an incremental increase in the number of tourists arriving in the country, up from 326 in May to 1,895 in August 2020.[16]

SWOC analysis of the tourism sector

The tourism sector has its own strengths, weaknesses, opportunities and challenges for engaging in the blue economy. The SWOC analysis presents the position that Tanzania Ministry of Tourism and Natural Resources showing the potential opportunities for engaging in the blue economy.

Recommendations for making Tanzania's tourism sector more 'blue'

Despite efforts made by the government to support the blue economy, including the implementation of a variety of policies (discussed above), there is a need to reactivate and promote the blue economy in practical terms. This can be achieved through holistic multi-sectoral planning and coordination of key sectors of the blue economy.

Effective engagement of the private sector is required for the sustainable management of the coastal and marine resources by developing and implementing cross-cutting policies and strategies, such as the Integrated Coastal Environment Management Strategy in Zanzibar and mainland Tanzania.

If the countries that border the Indian Ocean are to exploit the vast potential of the blue economy they need to uphold the IORA Plan of Action which has identified tourism and cultural exchanges within the Indian Ocean region as a strategic priority. Tanzania needs to adopt a strategic plan for the

Table 9.6 SWOC analysis of the tourism sector

Strengths	*Weaknesses*
Ocean and coastal tourism can create jobs and economic growth Tourism can be used as a tool to directly and indirectly support coastal and biodiversity conservation so that natural resources are protected for the long-term sustainability of the tourism sector and the economy Tailor-made product delivery Natural and cultural attraction resources	Lack of economies of scale Seasonality Environmental sustainability
Opportunities	*Challenges*
Coastal least developed countries and SIDS receive more than 41 million visitors per year Fishing tourism External market access to the EU, the USA, China and Russia Local tourism Agritourism	Environmental degradation Illegal fishing Pollution Climate change Concentration and globalization Global economic downfalls Health outbreaks such as COVID-19

Source: Author's compilation.

development of its coastal tourism over the the short, medium and long term.

The transportation sector

The transportation sector is another key economic activity operating marine and inland (road and railway) links to the blue economy under the Ministry of Works, Transport and Communication. The major blue economy related activities include import and export of Tanzania goods. The market share of ports to the blue economy contributes to the cargo transshipment for neighbouring land-linked countries. The upstream market in the transportation sector is organized through the importation of bulk dry and liquid goods as well as export of raw and semi-processed material goods primarily to markets in Europe, North America, China and Russia.

Marine transportation services

Marine transportation is offered by local and international shipping and marine transportation corporations. The major shipping lines operating in the Tanzanian ocean waters include Maersk shipping line with the largest fleet, followed by Mediterranean Shipping Company and Compagnie Générale Maritime.[17] The Dar es Salaam port is a major gateway for marine imports and exports with a handling capacity of 10.1 million metric tons per year.[18]

Tanzania Ports Authority

The Tanzania Ports Authority is responsible for the ports of Dar es Salaam, Tanga and Mtwara that offer marine transportation services for the import and export of goods destined for Tanzania and its land-linked neighbours.

In 2018/19 the Tanzania Ports Authority handled 6,934 million metric tons of goods, in comparison to 7,137 million tons in 2019/20. Meanwhile, the port of Dar es Salaam handled 4,694 million tons of goods in 2018/19 in comparison to 5,545 million tons of goods that were in transit to Uganda, Rwanda, Burundi, the Democratic Republic of the Congo, Zambia and Malawi.[19] Other countries that trade goods by traversing Tanzania's ocean waters include Zimbabwe, Comoro, South Sudan and Mozambique.[20]

Tanzania Shipping Agencies Corporation (TASAC) is another public institution established under Section 4 of the Tanzania Shipping Agencies Act No. 14 of 2017 that is beneficial to the blue economy. TASAC is the a legally authorized maritime administration for the country (section 11 of cap. 415), a regulator of maritime transport services (section 12 of cap. 415) and a maritime service provider (section 7 of cap. 415). Between 2018 and 2020 TASAC handled 687 vessels and issued 8,940 certificates to seafarers. The major global shipping lines serviced for carrying imports and exports to

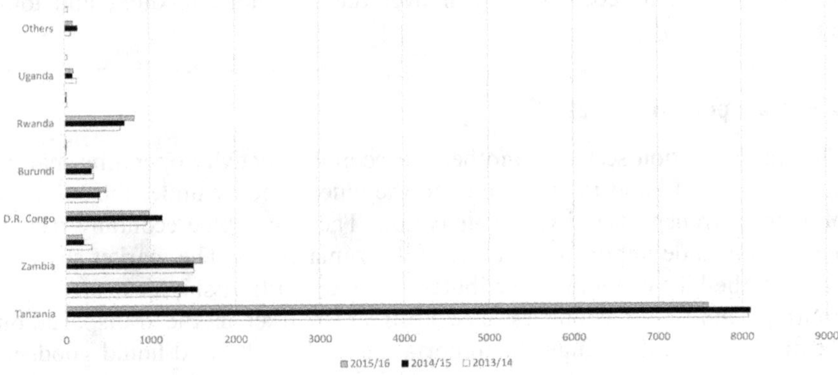

Figure 9.1 Import-export trade in the Indian Ocean
Source: Tanzania Ports Authority, http://ports.go.tz/index.php/en/publications/reports-annual-reports.

Tanzania include Maersk, Safmarine, MSC, Evergreen, Messina, Diamond Shipping, PIL, CMA-CGM and WEC Line.

Making Tanzania's transportation sector more 'blue'

Despite efforts by the government to improve ports, roads and railways that support the blue economy, marine transportation in the Indian Ocean is not well developed. There is a need to increase investment (public and private) in the country's cruise ships. AZAM Marine is a private shipping company that has introduced a ferry link between Dar es Salaam and Zanzibar that has revolutionized and modernized the sector, thus contributing to the blue economy in terms of goods and services as well as tourism. The service sector can be expanded to marine transportation between Dar es Salaam and Bagamoyo, Tanga, Mafia and Mtwara where cultural and heritage touristic sites exist. Similarly, the number of cruise ships sailing between Dar es Salaam and other IORA member countries for tourism purposes could be expanded, thus contributing to the blue economy.

Industries and international trade sectors

The Ministry of Industry, Trade and Investment has a central role in developing Tanzania's blue economy. Although the ministry is not directly involved in blue economy operations, it is mandated to manage domestic, international and intercontinental trade as well as managing and coordinating industries that rely upon oceanic waters for their trade. The ministry should focus its efforts on boosting processing and the addition of value, and on marine exports and the ship building industry, which has promising potential for job creation along the Indian Ocean. Although trade along and across the Indian Ocean is a historical phenomenon in East Africa ports, the modernization of Tanzania's ports and harbours could facilitate this.

Exports

Tanzania's chief exports include cotton seeds, sugarcane, uncoated craft paper, paperboard, sunflower seeds, maize (corn), black tea, minerals, cement clinkers and dried fish. Other export commodities are mattress and bedding linen, indigenous fabrics (khanga and kitenge) and livestock (mostly cattle and goats). In addition, Tanzania exports cashew nuts, tropical wood, Tanzanite, hides and skins, unwrought gold, pigeon peas, chickpeas, beans and textiles.

Imports

Tanzania imports goods such as medications, organic surface-active products, chemicals, vehicles and machinery. Other imported goods include iron and steel, oil and refined petroleum, fishing vessels, factory ships and processing vessels, yachts and vessels for pleasure or water sports, and rowing boats. In addition, Tanzania imports gas oil, kerosene-type jet fuel, pharmaceuticals, vaccines for human use, motorcycles with a reciprocating engine of capacity 50cc–250cc, sugar for industrial use and sodium hydroxide.

Table 9.7 shows the volume of imports and exports going through Tanzania's ports. The geographical position of Tanzania along the Indian Ocean coastline provides an ideal enabling environment for developing marine transportation for the blue economy. Tanzania has invested heavily in developing its ports of Dar es Salaam, Tanga and Mtwara that promises a significant contribution to the blue economy.

Tanzania's proximity to the Indian Ocean makes it a key partner to facilitate trade between Tanzania and other IORA member countries. Its main partners are Indonesia, the Comoros, India, and Indonesia. Figure 9.2 illustrates Tanzania's ocean-borne trade with Indonesia.

Table 9.8 illustrates the volume of trade in goods between Tanzania and the Comoros via the Indian Ocean.

Table 9.9 illustrates the volume of trade in goods between Tanzania and India via the Indian Ocean.

Conclusion: challenges and recommendations for the blue economy

This chapter puts forward proposals for the development and management of Tanzania's blue economy for the benefit of its inhabitants. Five key sectors were explored to find out how the current situation provides opportunities for the blue economy. As yet Tanzania does not have an exclusive, comprehensive policy for developing and managing its blue economy. Since the blue economy cuts across various issues, it currently operates using sectoral

Table 9.7 Tanzania's ocean-borne trade, 2009–19 (in US $ million)

Trade area	*Description*	*2009*	*2010*	*2011*	*2012*	*2013*	*2014*	*2015*	*2016*	*2017*	*2018*	*2019*
Europe	Exports	440.00	464.00	553.50	745.40	898.40	791.70	708.70	236.50	441.40	497.82	399.30
	Imports	1,076.60	1,111.40	1,074.40	1,536.90	2,759.40	2,895.00	1,159.80	557.70	936.10	1,015.76	909.01
	Total	1,516.60	1,575.40	1,627.90	2,282.30	3,657.80	3,686.70	1,868.50	794.20	1,377.50	1,513.57	1,308.31
	Balance of trade	–636.60	– 647.40	– 520.90	– 791.50	– 1,861.00	–2,103.30	– 451.10	– 321.20	– 494.70	– 517.94	– 509.71
EAC	Exports	263.80	450.10	352.40	515.30	419.10	598.10	1,062.40	437.70	349.60	447.50	674.40
	Imports	310.50	285.20	263.80	668.40	394.70	706.00	322.80	298.90	220.40	302.93	329.10
	Total	574.30	735.30	616.20	1,183.70	813.80	1,304.50	1,385.20	736.60	570.00	750.43	1,003.50
	Balance of trade	–46.70	164.90	88.60	– 153.10	24.40	– 108.30	739.60	138.80	129.20	144.57	345.30
SADC	Exports	374.20	625.10	1,158.80	1,421.90	1,243.55	1,235.90	1,357.70	1,017.90	877.80	999.34	1,330.90
	Imports	733.20	827.70	881.30	1,093.10	835.90	773.00	771.20	612.40	600.64	604.32	155.10
	Total	1,107.40	1,452.80	2,040.10	2,515.00	2,079.45	2,008.90	2,128.90	1,630.30	1,478.44	1,603.66	1,486.00
	Balance of trade	–359.00	–202.60	277.50	328.80	407.65	462.90	586.50	405.50	277.16	395.03	1,175.80

Source: Author's interview with Ministry of Industry, Trade and Investment officials, 2020.

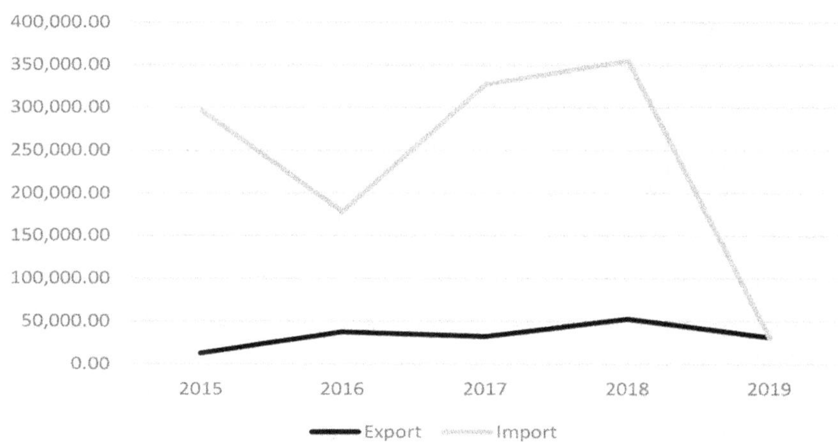

Figure 9.2 Tanzania's ocean-borne trade with Indonesia
Source: http://ports.go.tz/index.php/en/publications/reports-annual-reports.

ministry policies such as the National Environmental Policy for Environment Protection, the Fisheries Act for the Ministry of Livestock and Fisheries, and the Ministry of Tourism and Natural Resources Act for ocean resources. Consequently, the blue economy requires a standalone policy that will call for comprehensive implementation, monitoring and evaluation. Managing the blue economy requires the integration of science and technology innovations and environmental adaptations for the social well-being of all of the country's inhabitants.

The Indian Ocean is rich in resources including fish, oil and gas and a guiding policy is required to harness these resources in a sustainable manner. In this respect, Tanzania has done well by putting in place a national natural gas policy. Nevertheless, the missing link is a policy for all other sectors for harnessing the potential of the country's under- and unexploited marine resources. There is much more untapped potential across the sectors of transportation, tourism, energy, industries and fisheries. The IORA is the right platform for the member countries to collaborate on releasing the potential of Tanzania's blue economy. Policymakers need to increase support for research, science and technology innovations as well as developing public-private partnerships that can work for the national good.

In addition, multisectoral coordination to create an institutional, legal and strategic development framework is essential. Currently, the blue economy in Tanzania suffers from policy incoherence, weak enforcement and legislative gaps. The absence of multisectoral collaboration is hindering development.

Tanzania has not fully exploited the potential to expand the blue economy through the use of the abundant fisheries resources found in the waters of the Indian Ocean. The fisheries sector provides direct employment to 203,800 fishermen and women living along Tanzania's coastline. As the fisheries sector

Table 9.8 Trade statistics between Tanzania and the Comoros (amount in '000 million Tanzanian shillings)

Year	*2010*	*2011*	*2012*	*2013*	*2014*	*2015*	*2016*	*2017*	*2018*	*2019*
Exports	59,065.04	1,249.65	80.4	1,353.13	2,682.63	371,798.80	375,192.80	11,386.00	22,824.90	10,425.8
Imports	61.23	33.67	0.19	35.31	103.3	186.60	10,153.60	22.60	67.00	8.2
Trade volume	59,126.27	1,283.33	80.59	1,388.43	2,785.97	371,985.40	385,346.40	11,408.60	22,891.90	10,434
Balance of trade	59,003.8	1,215.97	80.21	1,317.81	2,579.32	371,612.20	365,039.20	11,363.40	22,757.90	10,417.6

Source: Author's interviews with Ministry of Industry, Trade and Investment, 2020.

Table 9.9 Trade statistics between Tanzania and India (amount in million US $)

Year	*2008*	*2009*	*2010*	*2011*	*2012*	*2013*	*2014*	*2015*	*2016*	*2017*	*2018*	*2019*
Exports	223.5	363.9	634.2	659.2	520.4	307.8	683.9	645.9	355.9	142.3	144.28	233.6
Imports	711.8	678.1	846.8	787.6	1,145.40	1,444.20	1,571.10	2,147.60	1,630.20	1,408.10	1,762.78	1,987.60
Trade volume	935.3	1,042.00	481.0	1,446.80	1,665.80	1,752.00	2,255.00	2,793.50	1,986.10	1,550.40	1,907.06	2,221.20
Trade balance	–488.3	–314.2	–212.6	–128.4	–625	–1,136.4	–887.2	–1,501.7	–1,274.3	–1,265.8	–1,618.50	–1,754.00

Source: Author's interviews with Ministry of Industry, Trade and Investment officials, 2020.

contributes 2.5% to GDP,[21] this sector needs a policy push to strengthen it. This chapter therefore recommends the formulation of a National Blue Economy Policy that will guide the relevant sectors in the harnessing the valuable resources found in the ocean and coastal zone. That policy should guide research, scientific and technological innovations, promote stakeholders' engagement in the blue economy and develop databases of socio-economic activities in the Indian Ocean so as to generate evidence-based policies.

This chapter has provided evidence of the important role played by the blue economy in Tanzania by utilizing the country's geographical location along the east coast of Africa. The trade and industries sector could do more to exploit the potential of the Indian Ocean. Knowledge and expertise exchange on the blue economy and its sustainability among stakeholders within the country as well as among IORA member countries has not been fully exploited. It is recommended that IORA should exploit the available expertise among its member countries to establish a number of vital related industries including fertilizers, natural gas, fish processing, ship building, chemicals and pharmaceuticals.

Oil refining and pumping has potential for developing the blue economy in Tanzania. In 2020 it was reported that TAZAMA (Tanzania Zambia Mafuta) Pipelines, a joint venture between the governments of Tanzania and Zambia, was seeking a loan to upgrade its 1,700 km-long Tanzania–Zambia pipeline that carries oil from Dar es Salaam to landlocked Zambia. This venture is expected to generate new jobs and revenue. A similar project is under consideration between Tanzania and Uganda to build an oil pipeline from Hoima in Uganda following the recent oil discovery there to Tanga port in Tanzania. Other similar opportunities may exist to pump natural gas and oil to neighbouring countries.

In conclusion, managing the blue economy is of paramount importance to Tanzania's socio-economic development. Since in the globalized world no country can achieve its development agenda alone, the Tanzanian government, in partnership with the IORA member states, should make concerted efforts to support the development of a national infrastructure, policy and regulations for harnessing marine resources, research and innovation, as well as the creation of network forums for knowledge sharing in the region.

Notes

1 www.iora.net.
2 Kathijotes (2013).
3 Government of Tanzania (2016). Annual Statistics Fisheries Report, Ministry of Livestock and Fisheries, Dar es Salaam, Tanzania.
4 Ibid.
5 Johannessen *et al.* (2018).
6 FISH-i-Africa (2019). https://fish-i-network.org/.
7 https://stopillegalfishing.com.
8 Ibid.

9 Christensen (2016).
10 Fronklin *et al.* (2012).
11 Government of Tanzania (2013).
12 Bishonge *et al.* (2018).
13 Government of Tanzania (2013).
14 Attri and Muller (2018).
15 Government of Tanzania (1989).
16 Author's interviews with Ministry of Natural Resources and Tourism officials, 2020.
17 Fitch Solutions (2016).
18 https://fish-i-network.org/.
19 Government of Tanzania (2020).
20 Ibid.
21 Bank of Tanzania (2020).

References

African Union (2012). *2050 Africa's Integrated Maritime Strategy.* Addis Ababa: African Union.

Attri, V. and Muller, N. (2018). *The Blue Economy Handbook of the Indian Ocean Region*. Pretoria: Africa Institute of South Africa.

Bank of Tanzania (2020). *Quarterly Economic Bulletin*, 52, No. 2 (June). Dar es Salaam: Bank of Tanzania.

Bishonge, O.K.*et al.* (2018). 'An Overview of the Natural Gas Sector in Tanzania: Achievements and Challenges', *Journal of Applied and Advanced Research*, 3(4).

BMIResearch FitchGroup. (2016), *Tanzania Shipping Report*. www.bmiresearch.com (accessed 31 August 2020).

Costanza, R.*et al.* (1999). 'Ecological Economics and Sustainable Governance of the Oceans', *Ecological Economics*, 31(2).

Christensen, J. (2016). 'Illegal, Unreported and Unregulated Fishing in Historical Perspective'. In K. Schwerdtner Máñez and B. Poulsen (eds), *Perspectives on Oceans Past: A Handbook on Marine Environmental History.* Dordrecht: Springer.

Economic Commission for Africa (UNECA) (2016). *Africa's Blue Economy: A Policy Handbook*. Addis Ababa: ECA.

European Commission (2012) *Blue Growth: Opportunities for Marine and Maritime Sustainable Growth.* Brussels: European Commission.

Fronklin, S.*et al.* (2012). 'Seaweed Mariculture as a Development Project in Zanzibar, East Africa: A Price too High to Pay?' *Aquaculture Journal*, 456.

Government of Tanzania (n.d.). *Tanzania Coastal and Marine Resources.* Dar es Salaam: Government of Tanzania.

Government of Tanzania (1989). *Territorial Sea and Exclusive Economic Zone Act*. Dar es Salaam: Government of Tanzania.

Government of Tanzania (1998). *The Deep Sea Fishing Authority Act*, Dar es Salaam: Government of Tanzania.

Government of Tanzania (2013). *The National Natural Gas Policy of Tanzania*. Dar es Salaam: Government of Tanzania, Ministry of Energy and Minerals.

Government of Tanzania (2016). *Annual Statistics Fisheries Report*. Dar es Salaam: Government of Tanzania, Ministry of Livestock and Fish Development.

Government of Tanzania (2018). *Tourism Statistical Bulletin Tourism Division*, Dodoma: Government of Tanzania.

Government of Tanzania (2020), *The Deep Sea Fisheries Management and Development Act*. Dodoma: Government of Tanzania.

Johannesen, T.*et al.* (2018). *Survey of Regional Resources and Ecosystem off South East Africa*. Bergen: Institute of Marine Research.

Kathijotes, N. (2013). 'Keynote: Blue Economy: Environmental and Behavioural Aspects Towards Sustainable Coastal Development', *Procedia: Social and Behavioral Sciences*, 101, pp. 7–13.

Lirasan, T. and Twide, P. (1993) 'Farming Eucheuma in Zanzibar, Tanzania'. In: A. R. O. Chapman, M. T. Brown and M. Lahaye (eds), *Fourteenth International Seaweed Symposium, Developments in Hydrobiology*, 85. Dordrecht: Springer.

Tanzania Coastal Management Partnership Support Unit and Mariculture Working Group (1999). *Tanzania Mariculture Issue Profile*, Working Document No. 5009. Dar es Salaam: TCMP.

Part III

Financing, measuring and governing the blue economy

10 Innovative financing for Africa's blue economy

Torsten Thiele, MPhil

Introduction

As demand for resources continues to grow and land-based sources come under strain, expectations for the ocean as a contributor to human development are increasing.[1] Yet at the same time, the health of the ocean, central to human well-being, is affected and may be reaching critical tipping points. Most fish stocks are overexploited, partly due to subsidies,[2] climate change and increased dissolved carbon dioxide which are changing ocean chemistry and disrupting species within food webs, and the fundamental capacity of the ocean to regulate the climate is being altered.[3] In order to address these challenges, innovative finance mechanisms based on sustainable finance approaches have been suggested as one way to build a different and more sustainable ocean economy,[4] generally referred to as the blue economy.[5]

The concept envisions a sustainable ocean-based economic model[6] that maintains coastal and marine ecosystems and resources. It employs environmentally sound and innovative infrastructure, technologies and practices, as well as institutional and financing arrangements, to meet the goals of sustainable and inclusive development; protecting coasts and oceans and reducing environmental risks and ecological scarcities; addressing water, energy and food security;[7] protecting the health, livelihoods and welfare of the people in the coastal zone; and fostering ecosystem-based climate change mitigation and adaptation measures.[8]

The transition to a sustainable blue economy is of particular importance to the coastal countries of sub-Saharan Africa[9] since their marine and coastal zones contribute a significant part of their food supply, livelihoods and well-being.[10] According to the Intergovernmental Science-Policy Platform on Biodiversity and Ecosystem Services (IPBES) report, in West Africa alone the coastal fisheries sector value added per year is US $4 billion, water purification services can be valued at $40,000 per sq km per year, mangrove coastal protection services at $4,500 per sq km per year[11] and coastal carbon sequestration services an average of $2,800 per sq km per year.[12] At the same time, West Africa's coasts are losing over $3.8 billion acres year to erosion, flooding and pollution.[13] Unsurprisingly, African business and political leaders are increasingly engaging with blue economy concepts.[14] This chapter examines the blue economy concept and discusses the increasing and cumulative challenges that require integrated solutions and finance approaches. It then describes the broader sustainable finance landscape before identifying a number of specific innovative finance approaches that can be applied to develop a sustainable blue economy. Finally, it proposes a number of governance mechanisms and other related interventions that will be beneficial to create an ocean finance architecture that is able to deliver the required transition to a blue economy at the necessary speed and scale.

Increasing and cumulative ocean challenges

According to the latest report of the Intergovernmental Panel on Climate Change that was set up by the World Meteorological Organization and the United Nations (UN),[15] detrimental changes are already occurring in the ocean and may be irreversible. Deoxygenation is leading to marine dead zones, acidification effects the growth prospects of marine organisms and warming shifts species' habitats and ecosystem patterns. Fisheries are already acutely affected.[16] Too little is being done to build resilient coastal infrastructure.[17] These issues need to be addressed as a matter of urgency[18] in order to avert a potential ecological disaster, requiring an ambitious agenda for global governance, including a robust High Seas Treaty. This new implementing agreement will not only provide an opportunity to implement appropriate area-based biodiversity protection measures for the global ocean, it will also regulate benefit-sharing for marine genetic resources.[19]

The Global Commission on Adaptation recommends a worldwide investment of US $1.8 trillion between 2020 and 2030 in early warning systems, climate-resilient infrastructure, improved dry land agricultural production, and global mangrove protection for an estimated $7.1 trillion in net benefits.[20] Marine energy holds significant potential to contribute to meeting the energy needs of the blue economy, with markets estimated in the billions of dollars.[21] All this will require new funding sources and financing mechanisms. Global warming is the pre-eminent factor driving change in the ocean but countries also need to rapidly address local stressors, such as overfishing and marine pollution.[22] The current benefits from the ocean economy are very unequally shared as a result of historical legacies and prevailing norms, particularly for women and indigenous peoples. This has brought global environmental challenges and negative effects on human well-being.[23] Finance has an important role to play to support both the roll-out of ocean solutions and to help to address the underlying causes and interdependencies.

The concept of innovative finance for the blue economy

The political declaration, 'Our ocean, our future: call for action' was adopted by the UN General Assembly on 6 July 2017. It recognizes the oceans 'as an engine for sustainable economic development and growth' and calls upon all stakeholders to conserve and sustainably use the oceans, seas and marine resources for sustainable development by taking actions to support, promote and strengthen sustainable ocean-based economies.[24] The blue economy has become an increasingly popular concept as a potential strategy for safeguarding the world's oceans and water resources.[25]

A recent European Parliamentary Research Service report[26] suggests that traditional economic activities relating to oceans and seas account for only 1.3% of the European Union (EU)'s gross domestic product (GDP), and that since 2009 the gross value of these established sectors, including fisheries, aquaculture, coastal tourism, maritime transport, port activities, shipbuilding and marine extraction of oil and gas, had increased by just 8% per annum. In contrast, a detailed World Bank study[27] states that 'doubling Mauritius' ocean economy from at present already about 10% of GNP [gross national product] is possible and worthwhile'. The study describes in detail how a significant investment programme, in sustainable port infrastructure and telecommunications services for instance, is financially viable and will to help to strengthen the island's economy in the longer term.

Innovative finance, the use of new finance mechanisms and broader sources of funding can help to deliver marine-related projects effectively.[28] Such approaches are consistent with long-term sustainability that can help to foster the transition to a vibrant blue economy that delivers opportunities, health, jobs and well-being. Without this strategy and supporting actions the ocean economy in a business-as-usual scenario will struggle as a consequence of the ongoing detrimental changes to ocean ecosystems from local and global stressors. A recent roadmap outlines these divergent futures in stark terms,[29] showing a trend for dwindling value and stranded assets in traditional extractive industries, while renewable, sustainable marine sectors such as offshore wind and marine biotechnology offer opportunities, jobs and other economic societal benefits.

A broad investor survey[30] has shown that interest in sustainable blue economy investments is high but industry expertise is low. According to the survey, there are opportunities for investment in early stage, impact and fixed income investments right now. Investments in sustainable blue economy infrastructure[31] and in listed companies in the areas are identified as future opportunities. Yet three out of four investors have not assessed their portfolios for their impact on the ocean which is a prerequisite for any informed decision-making. However, leading investors such as the Norwegian Sovereign Wealth Fund have already provided clear guidance to investee companies in the form of an ocean sustainability expectations document.[32]

A number of innovative finance instruments have already been used. Debt swaps use foreign grant support to transform existing sovereign debt into a commitment to deliver nature protection measures and have been applied in Seychelles.[33] Trust funds that directly provide donor money and technical assistance to deliver capacity building, for instance, are widely used by bilateral and multilateral donors in sub-Saharan Africa. Payments for ecosystem services such as through access to licenses and visitor fees or through wetland mitigation carbon credits are other examples.[34] These are some of the many mechanisms that can assist coastal sub-Saharan African developing countries to build appropriate investment frameworks for key sectors such as marine protection, plastics and waste management, maritime safety, aquaculture development and coastal infrastructure, including port logistics and offshore wind farms. Financial products such as 'blue bonds' and risk insurance can speed up the delivery of the money needed to undertake upfront capital expenditures and facilitate the participation of commercial private sector investors in this area. Blended finance[35] approaches which integrate a grant element into the transaction and therefore require a lower rate of financial return can make these transactions more affordable and can bring relevant expertise, credibility and verification opportunities to this process.

Sustainable finance

According to the recent 'Action Plan on Sustainable Finance' published by the European Banking Authority,

> Sustainable Finance can be broadly understood as financing and related institutional and market arrangements that contribute to the achievement of strong, sustainable, balanced and inclusive growth, through supporting directly and indirectly the framework of the Sustainable Development Goals (SDGs). This includes taking due account of environmental and social considerations in investment decision-making, leading to increased investments in longer-term and sustainable activities. It also includes methodologies for the assessment of the effective riskiness of exposures related to assets and activities associated substantially with environmental and/or social objectives.[36]

A wide range of concepts, products and approaches is needed to help the finance sector to make the transition to a long-term sustainable future. These include aligning the financial sector (in particular the multilateral development banks[37]) with the vision of the Paris Agreement, namely financing nationally determined contributions[38] and other forms of climate finance[39] and green and blue bonds;[40] also impact investments which are investments that are not only financially viable but are aligned with, verified and monitored for their delivery of pre-agreed goals such as specific environmental impacts (e.g. under the Blue Natural Capital Net Positive Impact Framework[41]), and payments for ecosystem services[42] and other ways to make efforts to achieve climate mitigation, adaptation and resilience. This can be through voluntary sales and market mechanisms, or pooled risk and insurance products. A relevant example is the Inter-American Development Bank's Framework and Principles for Climate Resilience Metrics in

Financing Operations[43] which provides targets to define and report on the contribution of financing activities directed towards climate-resilience objectives.

The sustainable finance sector has already had a significant history[44] and is rapidly developing. It is based on an emerging taxonomy of agreed principles and regulations, such as the green bond taxonomy proposed by the Technical Expert Group for Sustainable Finance.[45] The taxonomy sets performance thresholds (referred to as 'technical screening criteria') for economic activities which make a substantial contribution to one of six environmental objectives; do no significant harm (to the other five, where relevant); and meet minimum safeguards (e.g. the Organisation for Economic Co-operation and Development Guidelines on Multinational Enterprises and the UN Guiding Principles on Business and Human Rights).[46] Any innovative finance mechanisms for the blue economy need to be informed by and aligned with these important developments which will set the standards for the types of products that will be traded in the capital markets in the future. In addition to a product taxonomy the focus must be on protecting nature. Natural capital finance[47] and nature-based solutions are important aspects of this effort and need to be reflected in the blue economy context.

Innovative finance is based on the premise that traditional funding from taxpayers of public goods and delivery entirely through public services alongside purely commercial finance for private business are on their own insufficient to deliver the required transition at the necessary scale and speed and level of technology. Public-private partnerships and blended finance approaches, philanthropic support,[48] adoption of the sustainable blue economy finance principles[49] and other forms of targeted, cooperative approaches across multiple stakeholders will also be required.

Blue finance concepts

Frameworks such as the Sustainable Blue Economy Finance Principles[50] can help to transform traditional marine sectors as well as land-based activities that impact the ocean leading to a sustainable, resilient and circular economy whereby all resources are fully reused. In order to better engage other key industry sectors such as retail, health and financial services with the ocean and to create new opportunities and markets such as in bioproducts and data, ecosystem restoration and resilience investments will need ingenuity, time and money from many sources and through multiple avenues. Yet, as the World Bank study on Mauritius showed, the requisite investments are large but achievable. The study indicates that with a cumulative investment of US $5.8 billion over a 10-year period, the ocean economy's share of GDP should increase by almost 40% (rising from 12.6% to 17.5% of total GDP), and by more than 60% in absolute value.[51] Key strategies to address these challenges include stronger integration of sciences and ocean-observing systems; improved science-policy interfaces; new partnerships supported by a new ocean-climate finance system;[52] and improved ocean literacy and education to modify social norms and behaviours.[53]

By using comprehensive financing mechanisms for ocean management and coastal protection, including through nature-based solutions, and scaling up ocean science countries can put in place critical components to build resilience.[54] The Responsible Investor Survey[55] reported that

> the sectors believed [to offer the] best investment opportunities included climate change mitigation and adaptation (marine renewables), tackling marine plastic (and other) pollution, alongside sustainable fisheries and aquaculture but that there existed an urgent need to strengthen enabling conditions and develop innovative finance approaches to reduce risk by creating more sustainable projects with track records, fostering Public Private Partnerships and scaling investment using innovative finance approaches, such as blended finance.

Interdisciplinary and multidisciplinary research is vital in the discussion of the challenges and opportunities for blue growth.[56] This requires research into sectoral and regional approaches,[57] as the example of how to address harmful algae blooms in the EU shows.[58] Key sectors such as fisheries[59] need broader reform[60] that fully reflect changes in ecosystems under climate change and that strictly enforce scientific limits to overfishing. Additionally, blended finance[61] and impact investment strategies have to be engaged in order to put sectors such as fishing,[62] shipping and tourism onto the path towards sustainability. Examples include the full traceability of seafood, zero-carbon shipping and low-impact tourism. As pressures on the ocean mount, systematic social and ecological screening needs to become the norm for mainstream financial mechanisms (e.g. credit lending), in the same manner as is currently the case for financial auditing.[63] Larger efforts such as designing a global ocean data infrastructure for networked remote sensing can provide many countries and companies with critical information. Tools such as marine spatial planning[64] can help to optimize the use of the marine spaces, as the new marine spatial plan for Seychelles shows. Seychelles, with an exclusive economic zone of 1,374,000 sq km is leading the region in this effort. A full integration of the marine sphere[65] into the global climate finance architecture[66] is required if countries want to be able to appropriately use the instruments that land-based green finance has already developed.

As coastal areas are particularly affected by sea level rise and storm surges[67] investment in coastal resilience is required to address ocean risk.[68] In particular, countries will have to better integrate nature-based solutions into climate solutions[69] and infrastructure design. Such blue infrastructure efforts can, as a recent International Union for Conservation of Nature (IUCN) report suggests,[70] reduce risks and improve the financial considerations by delivering multiple benefits such as increased resilience, gradual adaptation and enhanced biodiversity. The concept of blue natural capital[71] not only provides a broader narrative around the multiple co-benefits of an activity that addresses not only a mitigation challenge[72] but helps to deliver local livelihoods through job opportunities[73] and to protect biodiversity. Over time it can develop a new and distinct asset class of comprehensively sustainable spaces. This requires appropriate processes to assess the net positive impact of any activities in the blue arena, for which IUCN has already worked on a methodology[74] which can be applied through the Blue Natural Capital Positive Impacts Framework.[75]

A report[76] for the High Level Panel for a Sustainable Ocean Economy[77] recently proposed multiple avenues for ocean-climate finance solutions. Ocean accounts as outlined in a recent Blue Paper[78] and additional financing tools covering the whole process from project preparation via project debt and equity to capital market instruments such as blue bonds,[79] as discussed above, all are needed to deliver investment in ocean knowledge, clean energy, water solutions, smart coastal infrastructure and the marine bio-economy, thereby securing the links between ocean health and human health. Blended finance approaches[80] will be critical and have already been successfully applied in various sectors,[81] such as energy and transport. Strengthening risk mitigation and insurance tools[82] and building regional hubs and accelerators could help to engage less developed countries in sub-Saharan Africa.

Conclusion

To address concerns about a rapid industrialization of the ocean, referred to as 'the blue acceleration,[83] countries need to speed up actions to support protection[84] and restoration[85] of blue carbon ecosystems[86] and other ocean and marine habitats. This also includes the high seas where a new biodiversity implementing agreement being negotiated at the UN[87] can help to address ocean governance challenges in light of the tragedy of the commons. Supportive actions along coasts, marine regions[88] and across sea basins require innovative ocean finance.[89] Marine and coastal blue natural capital assets[90] are underpinning a robust and resilient blue economy but they are under threat. Strong nationally determined contributions under the Paris Agreement[91] that cover ocean goals, marine spatial plans with ecosystem protection for resilience and biodiversity[92] that engage all stakeholders and adequate ecosystem-based management mandates[93] are a prerequisite to attracting long-term sustainable financing into sub-Saharan Africa's blue economy.[94]

Various avenues will be needed that show investors how blue carbon finance[95] (i.e. the integration of coastal and marine carbon capture into international carbon markets and other ecosystem-based investment funding) can be accessed by African coastal states in order to support blue carbon ecosystems, conservation and the appropriate restoration actions that will then be required. Accessing climate finance tools such as the Green Climate Fund as well as facilitating private investor engagement, with incentives for a rapid transition to clean energy and nature-based blue infrastructure, are ways to integrate innovative finance into a nation's blue economy strategy.

Multilateral development banks play a key role in delivering the substantial investments needed by developing countries in the field of coastal and maritime infrastructure.[96] For example, the PROBLUE trust fund established by the World Bank to help to manage marine litter in developing countries and to support transition to the blue economy and the realization of SDG 14 is supported by many donor countries. Similarly, the Asian Development Bank's US $5 billion Action Plan for Healthy Oceans, Sustainable Blue Economies[97] contains an oceans financing initiative that aims to create opportunities for the private sector to invest in projects that improve the ocean's health and stimulate the blue economy. This is particularly important for the blue economy because, as this chapter has stressed, ocean solutions require broad stakeholder engagement and an equitable approach. Moreover, maritime property rights are often less clearly defined than on land, with overlapping state rights and traditional usage in coastal areas.[98]

As the Indian Ocean Rim Association's Blue Carbon Hub inaugural think tank meeting on blue carbon finance held in 2020 in Mauritius showed,[99] involving regional organizations can help to boost knowledge sharing and capacity building. Asset owners and other private financiers can be engaged through an ocean risk[100] and sustainable natural capital investment focus. Science-based approaches, including by reference to the decade of ocean science[101] and innovative ocean technologies can deliver attractive investment propositions. The overall scale and capacity of the seas to rapidly grow biomass makes them a natural space for solutions to food and energy challenges but requires supportive finance and insurance[102] approaches. As the Blue Paper on food for the ocean panel suggests,[103] sustainably expanding the mariculture of species such as bivalves and species that do not depend on feed inputs for their nutrition can substantially increase the supply of nutritious food while simultaneously having a lower impact on the marine environment.

Ocean finance based on the Sustainable Blue Economy Finance Principles and a wider framework of an ocean finance architecture[104] is a crucial part of the transition to sustainability in sub-Saharan Africa. Globally, current natural climate solutions[105] efforts receive only 0.8% of public and private climate financing, despite offering roughly 37% of the potential mitigation needs through 2030.[106] Sub-Saharan African economies will need to proactively target the sustainable blue economy using innovative finance to build sustainable marine and coastal economies. Rebuilding marine life represents a critical challenge for humanity and a smart economic objective to achieve a sustainable future.[107] Just as the mobile telecommunications revolution has demonstrated how African operators have been able to create new solutions appropriate to their local conditions, smart innovative finance that is tailored to coastal communities in particular can also be a key contributor to the delivery of a vibrant and sustainable blue economy for the nations of sub-Saharan Africa.

Notes

1 Jouffray, J.-B., Blasiak, R., Norstroem, A.V. *et al.* (2020) 'The Blue Acceleration: The Trajectory of Human Expansion into the Ocean'. *One Earth: Perspective.* https://doi.org/10.1016/j.oneear.2019.12.016.

2 Sumaila, U. R., Ebrahim, N., Schuhbauer, A. *et al.* (2019). Updated Estimates and Analysis of Global Fisheries Subsidies'. *Marine Policy,* 109, https://doi.org/10.1016/j.marpol.2019.103695103695.

3 Claudet, J., Bopp, L, Cheung, W. W. L. *et al.* (2019) 'A Roadmap for Using the UN Decade of Ocean Science for Sustainable Development in Support of Science, Policy, and Action'. *OneEarth.*

4 Organisation for Economic Co-operation and Development (OECD) (2016) *The Ocean Economy in 2030*, OECD Publishing, Paris. https://doi.org/10.1787/9789264251724-en.

5 Michel, J. A. (2016) *Rethinking the Oceans: Towards the Blue Economy.* Paragon House, Trowbridge.

6 Ebarvia, M. C. M. (2016) 'Economic Assessment of Oceans for Sustainable Blue Economy Development'. *Journal of Ocean and Coastal Economics,* Vol. 2, Article 7. doi:h>p://dx.doi.org/10.15351/2373–8456.1051.

7 Froehlich H. E., Gentry, R. R. and Halpern, B. S. (2018). 'Global Change in Marine Aquaculture Production Potential under Climate Change. *Nature Eco Evo.*

8 Bennett, N. J., Cisneros-Montemayor, A. M., Blythe, J., Silver, J. J., Singh, G., Andrews, N., Calo, A., Christie, P., Di Franco, A., Finkbeiner, E. M. *et al.* (2019). 'Towards a Sustainable and Equitable Blue Economy'. *Nat. Sustain.* 2, 991–93.

9 See the Nairobi Statement of Intent on Advancing the Global Sustainable Blue Economy. www.blueeconomyconference.go.ke/wp-content/uploads/2018/11/Nairobi-Statement-of-Intent-Advancing-Global-Sustainable-Blue-Economy.pdf.

10 Intergovernmental Science-Policy Platform on Biodiversity and Ecosystem Services (IPBES) 'Biodiversity and nature's contributions continue dangerous decline, scientists warn: Human well-being at risk. Landmark reports highlight options to protect and restore nature and its vital contributions to people'. *ScienceDaily*, 23 March 2018. www.sciencedaily.com/releases/2018/03/180323093734.htm.

11 The total length of the African coastline is 30,500 km.

12 The report mentioned above also contains similar data for other African regions.

13 Croitoru, L., Miranda, J. J. and Sarraf, M. (2017) *The Cost of Coastal Zone Degradation in West Africa.* World Bank, Washington, DC.

14 www.choiseul-africa-businessforum.com.

15 IPCC (2019) 'Summary for Policymakers'. In H.-O. Pörtner, D. C. Roberts, V. Masson-Delmotte *et al.* (eds) *IPCC Special Report on the Ocean and Cryosphere in a Changing Climate.* IPCC, Geneva.

16 Barange, M., Bahri, T., Beveridge, M. C. M. *et al.* (eds) (2018) *Impacts of Climate Change on Fisheries and Aquaculture: Synthesis of Current Knowledge, Adaptation and Mitigation Options.* FAO Fisheries and Aquaculture Technical Paper No. 627. Rome, Food and Agricultural Organization of the United Nations, p. 628.

17 Thacker, S., Adshead, D., Fay, M. *et al.* (2019) 'Infrastructure for Sustainable Development'. *Nature Sustainability* 2(4).

18 Arico, S. (2015) 'Issues Regarding Oceans and Opportunities: An Introduction to the Book'. In S. Arico (ed.) *Ocean Sustainability in the 21st Century.* UNESCO Publishing, Paris.

19 Collins, J., Harden-Davies, H., Vanagt, T. *et al.* (2019) 'Inclusive Innovation: Enhancing Global Participation and Benefit Sharing Linked to the Utilization of Marine Genetic Resources from ABNJ'. *MarPol.*

20 Harou, J. J., Matthews, J. H., Smith, D. M. *et al.* (2020) 'Water at COP25: Resilience Enables Climate Change Adaptation Through Better Planning, Governance and Finance'. *Proc of the Institution of Civil Engineers – Water Management* 173(2): 55–58. https://doi.org/10.1680/jwama.173.2020.2.55.

21 LiVecchi, A., Copping, D. and Jenne, A. (2019) *Powering the Blue Economy; Exploring Opportunities for Marine Renewable Energy in Maritime Markets.* U.S. Department of Energy, Office of Energy Efficiency and Renewable Energy. Washington, DC.

22 Laffoley, D., Baxter J. M., Amon, D. J. *et al.* (2020) 'Urgent Steps Needed to Restore Ocean Health'. *Aquat. Conserv. Mar. Freshw. Ecosyst.* doi: 10.1002/aqc.3182.

23 Österblom, H., Wabnitz, C. C. C., Tladi, D. *et al.* (2019) *Towards Ocean Equity.* World Resources Institute, Washington, DC. www.oceanpanel.org/how-distribute-benefits-ocean-equitably.

24 https://enb.iisd.org/oceans/blueeconomy/2018/html/enbplus208num31e.html.

25 Lee, H. L, Junsung, N. and Jong, S. K. (2020) 'The Blue Economy and the United Nations' Sustainable Development Goals: Challenges and Opportunities'. *Environment International*, vol. 137, April.

26 Scholaert, F. *et al.* (2020) *The Blue Economy: Overview and EU Policy Framework.* European Parliamentary Research Service. PE 646.152 – January.

27 Cervigni, R. and Scandizzo, P. L. (eds) (2017) *The Ocean Economy in Mauritius.* World Bank, Washington, DC.

28 Thiele, T. and Gerber, L. R. (2017). 'Innovative Financing for the High Seas'. *Aquat. Conserv. Mar. Freshw. Ecosyst.* 27, 89–99.

29 See Figure 4 in Claudet *et al.* 'A Roadmap for Using the UN Decade of Ocean Science'.

30 *Investors and the Blue Economy* (2020) Responsible Investor & Credit Suisse. www.responsible-investor.com/reports/responsible-investor-and-credit-suisse-or-investors-and-the-blue-economy.

31 Thiele, T., Alleng, G., Biermann, A. *et al.* (2020). 'Towards Sustainable Blue Infrastructure Finance: The Need, Opportunity and Means to Integrate Nature-Based Solutions into Coastal Resilience Planning and Investments. Key Messages'. IUCN, Gland.

32 www.nbim.no/contentassets/17ed97a1a9f845ad8e847a51bc4b8141/nbim_expectations_oceans.pdf.

33 For the Seychelles example, see Silver, J. J. and Campbell, L. M. (2018) Conservation, Development and the Blue Frontier: The Republic of Seychelles' Debt Restructuring for Marine Conservation and Climate Adaptation Program, John Wiley & Sons. www.researchgate.net/profile/Jennifer_Silver/publication/327629338_Conservation_development_and_the_blue_frontier_the_Republic_of_Seychelles'_Debt_Restructuring_for_Marine_Conservation_and_Climate_Adaptation_Program/links/5b9e75b8299bf13e603677d8/Conservation-development-and-the-blue-frontier-the-Republic-of-Seychelles-Debt-Restructuring-for-Marine-Conservation-and-Climate-Adaptation-Program.pdf.

34 Spalding, M. D., Brumbaugh, R. D. and Laandis, E. (2016) *Atlas of Ocean Wealth.* The Nature Conservancy, Arlington, VA.

35 OECD/UNCDF (2019) *Blended Finance in the Least Developed Countries 2019*, OECD Publishing, Paris. https://doi.org/10.1787/1c142aae-en.

36 European Banking Authority (EBA) (2019) 'Action Plan on Sustainable Finance', 9 December, EBA, Paris.

37 https://newclimate.org/2020/03/19/multilateral-development-banks-have-made-progress-towards-paris-alignment-but-still-need-to-fill-in-critical-details/.

38 EBRD (2019) *Joint Report on Multilateral Development Banks' Climate Finance.* European Bank for Reconstruction and Development, London.

39 Climate Bonds Initiative (CBI) (2020) *2019 Green Bond Market Summary.* CBI, London. See www.climatebonds.net/resources/reports/2019-green-bond-market-summary (accessed 6 March 2020).

40 World Bank (2018) '10 Years of Green Bonds: Creating the Blueprint for Sustainability across Capital Markets', 18 March. World Bank, Washington, DC.

41 https://bluenaturalcapital.org/wp2018/wp-content/uploads/2019/03/BNC-Framework_final.pdf.

42 Schomers, S. and Mtzdorf, B. (2013) 'Payments for Ecosystem Services: A Review and Comparison of Developing and Industrialized Countries'. *Ecosystem Services*, vol. 6, pp. 16–30.

43 Inter-American Development Bank (IDB) (2019) *A Framework and Principles for Climate Resilience Metrics in Financing Operations.* IDB, Washington, DC. https://publications.iadb.org/en/framework-and-principles-climate-resilience-metrics-financing- operations (accessed 6 March 2020).

44 World Bank (2011) *Research Report on Leading Initiatives and Literature Related to Green Infrastructure Finance.* World Bank, Washington, DC.

45 *Taxonomy: Final Report of the Technical Expert Group on Sustainable Finance*, March 2020.

46 www.greenfinancelac.org/wp-content/uploads/2020/03/200309-sustainable-finance-teg-final-report-taxonomy_en.pdf.

47 Cooper, G. and Trémolet S. (2019) *Investing in Nature: Private Finance for Nature-Based Resilience.* The Nature Conservancy and Environmental Finance. London.

48 Wabnitz, C. and Blasiak, R. (2019). 'The Rapidly Changing World of Ocean Finance'. *Mar. Policy* 107.

49 Declaration of the Sustainable Blue Economy Finance Principles (2018). https://ec.europa.eu/maritimeaffairs/sites/maritimeaffairs/files/declaration-sustainable-blue-economy-finance-principles_en.pdf.

50 Ibid.

51 Cervigni, R. and Scandizzo, P. L. (eds) (2017) 'The Ocean Economy in Mauritius'. World Bank, Washington, DC.

52 Cicin-Sain, B. *et al.* (2019) *Assessing Progress on Ocean and Climate Action: A Report of the Roadmap to Oceans and Climate Action* (ROCA) Initiative. https://roca-initiative.com/reports/.

53 Claudet *et al.* (2019).

54 Laffoley, D., Baxter, J. M., Amon, D. J. *et al.* (2020) 'Urgent Steps Needed to Restore Ocean Health'. *Aquat. Conserv. Mar. Freshw. Ecosyst.* doi: 10.1002/aqc.3182.

55 EU Technical Expert Group on Sustainable Finance (2020) 'Taxonomy: Final Report of the Technical Expert Group on Sustainable Finance. *Climate ADAPT.*

56 Klinger, D. H., Eikeset, A. M. and Davidsdottir, B. (2018) 'What Is Blue Growth? The Semantics of "Sustainable Development" of Marine Environments'. *Mar Pol*, pp. 356–62. https://doi.org/10.1016/j.marpol.2017.10.019.

57 Kulkarni, R. (2018). 'Innovative Ffinancing and Regional Dialogue Are Central for a Thriving "'Blue Economy"'. Blog. UNDP.

58 Wenhai, L., Cusack, C., Baker, M. *et al.* (2019) 'Successful Blue Economy Examples with an Emphasis on International Perspectives'. *Front. Mar. Sci.*, 7 June. https://doi.org/10.3389/fmars.2019.00261.

59 de Graaf, G. and Garibaldi, L. (2014) 'The Value of African Fisheries. FAO Fisheries and Aquaculture Circular. No. 1093. Rome, FAO, p. 76.
60 Costello, C. *et al.* (2016) 'Global Fishery Prospects under Contrasting Management Regimes'. *Proc Natl Acad Sci*, 113, pp. 5125–29.
61 Environmental Defense Fund and Nicholas Institute for Environmental Policy Solutions at Duke University (2018). *Financing Fisheries Reform: Blended Capital Approaches in Support of Sustainable Wild-Capture Fisheries.* edf.org/blendedcapital.
62 Encourage Capital (2016) *Investing for Sustainable Global Fisheries.* http://investinvibrantoceans.org/wp-content/uploads/documents Nexus_Blue_Strategy_FINAL_1-11-16.pdf.
63 Jouffray, J.-B., Crona, B. and Wassenius, E. *et al.* (2019) 'Leverage Points in the Financial Sector for Seafood Sustainability. *Sci. Adv. 5.*
64 European Commission Study (2018) *Maritime Spatial Planning (MSP) for Blue Growth.* Technical Study. Publications Office of the European Union, Luxembourg.
65 Because the Ocean (2019) *Ocean for Climate: Ocean-Related Measures in Climate Strategies.* www.becausetheocean.org/ocean-for-climate/.
66 Thiele, T. (2019) 'Adding "Blue" to International Climate Finance'. In *From "Green" to "Blue Finance": Integrating the Ocean into the Global Climate*, LSE Global Policy Lab. www.lse.ac.uk/iga/assets/documents/global-policy-lab/From-Green-to-Blue-Finance.pdf.
67 Hendrick, G. (2020) 'Ocean Risk and Resilience', Jan.–Feb., *Leader's Edge Magazine.*
68 www.oceanriskalliance.org.
69 C2G (2019) *C2G Evidence Brief: Governing Nature-Based Solutions to Carbon Dioxide Removal.* Carnegie Climate Governance Initiative, New York. p. 27, Version 20190824. www.c2g2.net/wp-content/uploads/c2g_evidencebrief_NBS.pdf.
70 Thiele, T., Alleng, G., Biermann, A., Corwin, E., Crooks, S., Fieldhouse, P. *et al.* (2020) *Blue Infrastructure Finance: A New Approach, Integrating Nature-Based Solutions for Coastal Resilience.* IUCN, Gland.
71 Bluenaturalcapital.org.
72 Froehlich, H. E., J. C. Afflerbach, M. Frazier and B. S. Halpern (2019) 'Blue Growth Potential to Mitigate Climate Change through Seaweed Offsetting'. *Current Biology.* doi:10.1016/j.cub.2019.07.041.
73 van den Burg, S. W. K., Dagevos, H. and Helmes, R. J. K. (2019) 'Towards Sustainable European Seaweed Value Chains: A Triple P Perspective'. *ICES Journal of Marine Science*, doi:10.1093/icesjms/fsz183.
74 www.iucn.org/theme/business-and-biodiversity/our-work/business-approaches-and-tools/business-and-biodiversity-net-gain.
75 https://bluenaturalcapital.org/wp2018/wp-content/uploads/2019/03/BNC-Framework_final.pdf.
76 Hoegh-Guldberg. O., Caldeira, K., Chopin, T. *et al.* (2019) *The Ocean as a Solution to Climate Change: Five Opportunities for Action.* Report. World Resources Institute, Washington, DC. www.oceanpanel.org/climate.
77 www.oceanpanel.org/about-the-panel.
78 Fenichel, E. P., Milligan, B., Porras, I. *et al.* (2020) *National Accounting for the Ocean & Ocean Economy.* Blue Paper #8. World Resources Institute, Washington, DC.
79 Roth, N., Thiele, T. and von Unger, M. (2019) *Blue Bonds: Financing Resilience of Coastal Ecosystems.* A technical guideline prepared for IUCN GMPP. IUCN, Gland.
80 Rode, J., Pinzon, A., Stabile, M. C. C. *et al.* (2019) 'Why "Blended Finance" Could Help Transitions to Sustainable Landscapes: Lessons from the Unlocking Forest Finance Project'. *Science Direct*, p. 37.

81 Basile, I. and J. Dutra (2019) 'Blended Finance Funds and Facilities: 2018 Survey Results'. OECD Development Co-operation Working Papers, No. 59, OECD Publishing, Paris. https://doi.org/10.1787/806991a2-en.

82 Jarzabkowski P., Chalkias, K., Clarke, D. *et al.* (2019) *Insurance for Climate Adaptation: Opportunities and Limitations.* GCA, Rotterdam. https://cdn.gca.org/assets/2019–12/Insurance%20for%20climate% 20adaptation_Opportunities%20and%20Limitations.pdf.

83 Jouffray, J.-B., Blasiak, R., Norström, A. V. *et al.* (2019) 'The Blue Acceleration: The Trajectory of Human Expansion into the Ocean'. *One Earth.* doi:10.1016/j.oneear.2019.12.016.

84 Gjerde, K. M., Nordtvedt Reeve, L. L., Harden-Davies, H. *et al.* (2016) 'Protecting Earth's Last Conservation Frontier: Scientific, Management and Legal Priorities for MPAs beyond National Boundaries'. *Aquatic Conserv.*https://doi.org/10.1002/aqc.2646.

85 Layton, C., Coleman, M. A., Marzinelli, E. M. (2020) 'Kelp Forest Restoration in Australia'. *Front. Mar. Sci.*, 14 February. https://doi.org/10.3389/fmars.2020.00074.

86 Gattuso, J. P. (2019) Opportunities for Increasing Ocean Action in Climate Strategies. *Iddri Policy Brief* no. 02/19. http://bit.ly/2NswD9t.

87 www.iucn.org/sites/dev/files/content/documents/iucn_comments_on_revised_bbnj_draft_text_February_2020.pdf.

88 Gjerde, K. and Wright, G. (2019) 'Towards Ecosystem-based Management of the Global Ocean: Strengthening Regional Cooperation through a New Agreement for the Conservation and Sustainable Use of Marine Biodiversity in Areas Beyond National Jurisdiction', STRONG High Seas Project.

89 Walsh, M. (2018) Ocean Finance: Definition and Actions. Pacific Ocean Finance Program, Pacific Islands Forum Fisheries Agency and Office 6.

90 Thiele, T. (2019) 'Ocean Bleu: des risques lies a la dégradation de l'ocean a une assurance pour le patrimoine corallien. Concepts pour sauvegarder le capital natural bleu'. In D. Bretones, *Les Organisations face aux defies technologies et societaux due XXIe siecle.* MA Editions, Paris.

91 https://unfccc.int/process-and-meetings/the-paris-agreement/nationally-determined-contributions-ndcs.

92 OECD (2019) *Biodiversity: Finance and the Economic and Business Case for Action.* OECD, Paris. www.oecd.org/environment/resources/biodiversity/biodiversity-finance-and-the-economic-and-business-case-for-action.htm.

93 Link, J. S., Dickey-Collas, M., Rudd, M. *et al.* (2018) 'Clarifying Mandates for Marine Ecosystem-Based Management'. *ICES Journal of Marine Science.*

94 Thiele, T. (2015) 'Sauver l'océan, protéger la haute mer'. *Géoéconomie*, 76.

95 Gordon, D., Maurray, B. C., Pendleton, L. and Victor, B. (2011) 'Financing Options for Blue Carbon Opportunities and Lessons from the REDD+ Experience'. Nicholas Institute for Environmental Policy Solutions Report. NI R 11–11.

96 Thiele, T., Alleng, G., Biermann, A. *et al.* (2020) *Towards Sustainable Blue Infrastructure Finance: The Need, Opportunity and Means to Integrate Nature-Based Solutions into Coastal Resilience Planning And Investments. Key messages.* IUCN, Gland.

97 https://sdg.iisd.org/news/adb-launches-usd-5-billion-action-plan-for-healthy-oceans-sustainable-blue-economies/.

98 Hirsch, P. (2013) Strategy to Address Coastal Property Rights and Resource Tenure Issues for MFF Phase 3. www.mangrovesforthefuture.org/assets/Repository/Documents/130827-Hirsch-strategy-property-rights-and-resource-tenure-issues-for-2.pdf.

99 www.iora.int/en/events-media-news/events/priorities-focus-areas/blue-economy/2020/iora-indian-ocean-blue-carbon-hub-inaugural-think-tank-blue-carbon-finance.

100 Niehörster, F. and Murnane, R. J. (2018) *Ocean Risk and the Insurance Industry.* XL Catlin Services SE, May, 40 pp.

101 Claudet *et al.* (2019).

102 Better Insurance Is Vital to Protecting Our Ocean. https://chinadialogueocean.net/2760-better-insurance-is-vital-to-protecting-our-ocean/.

103 Costello, C., Cao, L., Gelcich, S. *et al.* (2019) *The Future of Food from the Sea.* World Resources Institute, Washington, DC. www.oceanpanel.org/future-food-sea.

104 Thiele (2019) 'Adding "Blue" to International Climate Finance'.

105 Griscom, B. W. Adams, J. Ellis, P. W. *et al.* (2017) 'Natural Climate Solutions'. *Proc. Natl. Acad. Sci.*, 114, 11645–650

106 Fargione, J. E., Bassett, S., Boucher, T. *et al.* (2018) 'Natural Climate Solutions for the United States'. *Science Advances. Environmental Studies.*

107 Duarte, C. M., Agusti, S., Barbier, E. *et al.* (2020) 'Rebuilding Marine Life'. *Nature* 580, pp. 39–51. https://doi.org/10.1038/s41586-020-2146-7.

11 Measuring the blue economy

Charles S. Colgan, PhD, Vivian Louis Forbes, PhD and Iddi Mwanyoka, MSc

Introduction

The concept of a 'blue economy' originated from the United Nations (UN) Conference on Sustainable Development (Rio+20) held in Rio de Janeiro, Brazil, in 2012. The focus of the conference was on two themes: further developing and enhancing the institutional framework for sustainable development; and advancing the 'green economy' concept, namely taking measures to achieve economic growth while protecting the environment. Poverty eradication was defined as a priority consistent within both objectives.[1] However, throughout the preparatory process for Rio+20, many coastal states and Small Island Developing States (SIDS) questioned the focus on the green economy alone and its applicability to them. These countries argued for a blue economy focus that would apply the same principles of the green economy to the economy of oceans and coasts.

The years since Rio+20 have seen the blue economy concept diffuse rapidly around the globe, with blue economy strategies emerging in many countries. Blue economy studies of one kind or another have been conducted by international organizations such as the UN and the World Bank[2] as well as various countries[3] and regions.[4] In Africa, oceans and the blue economy have received attention in a number of forums. The UN has issued country-specific policy guidance[5] as well as considering Africa-wide potential strategies for the blue economy.[6] The African Union directed its attention to blue economy issues as early as 2012 when it published a maritime strategy.[7] This was updated and expanded in 2019 to include a blue economy strategy for Africa.[8]

The increasing level of attention being given to the blue economy has come about alongside increased attention to the definition and assessment of the ocean economy within traditional measures. The ocean economy is briefly defined as that portion of national income, output and employment that is derived directly or indirectly from the oceans. The term ocean economy emerged in the United States, where the first attempt to define and measure the portion of the economy related to the oceans was made in 1974.[9] National and regional ocean economy measurements were developed for the

United States and have been continually produced since 2001.[10] Other countries began similar measurements of their own ocean-related economies, including Canada,[11] Australia,[12] and New Zealand.[13]

There is considerable confusion between the terms 'ocean economy' and 'blue economy'. The two terms are often used interchangeably, which is, in one way, positive because the use of the two terms offers many opportunities to raise awareness among governments and stakeholders of the importance of the ocean both as a key element of 'natural capital' and a driver of growth in national and regional economies. But there are some differences between the way in which the two terms have evolved that point to a clear separation of meaning with important implications for the future of ocean-related policies.

A somewhat simplified distinction between the 'ocean' and 'blue' can be described in two ways. The blue economy idea includes the ocean economy idea, but expands it to explicitly consider the environmental, ecological and social dimensions of economic uses of the ocean.[14]At the same time, the blue economy's broader perspective is widely shared but is not accompanied by the ocean economy's focus on measurement of the size and shape of the economy. Thus, the blue economy motivates action, but it is the data relating to the ocean economy that lays the foundation for choosing which actions to take and evaluating those choices. The challenge, then, is to merge the measurements of the ocean economy with the vision of the blue economy as a guide to a sustainable future.

Fortunately, the framework to accomplish this merger exists with developments in economic accounting led by the UN. This chapter uses these current and evolving frameworks for measuring economic value to build a comprehensive measurement of the blue economy. The chapter begins by describing how to construct ocean economy accounts, which provide foundational information for the blue economy. It then discusses how to incorporate broader environmental and ecological values. It concludes by considering limitations on the current approaches to measurement of the ocean economy and provides a course of action to create the measurements needed for a successful transition to a blue economy.

Measuring the ocean economy

Measuring the ocean economy is accomplished primarily by rearranging existing publicly collected data with additional information as needed. Most nations already have the essential ingredients needed to measure the ocean economy in their national accounting and economic statistics and this data is collected and maintained in accordance with the international System of National Accounts standards.[15] Unfortunately, existing systems are not usually organized to make the identification of ocean-related economic activity easy. Appropriately arranging even existing data can still require substantial effort.

The construction of a set of ocean accounts requires that a number of tactical decisions must be made. These can be summarized in the following checklist for an ocean economy data development project.

1 Top-down versus bottom-up
2 Industry
3 Geography
4 Economic concepts
5 Final and intermediate goods and services
6 Stocks versus flows
7 Past and future

Top-down versus bottom-up

There are two approaches to rearranging existing economic data systems to reflect the ocean relationship. One is to take existing industry-based data and identify some proportion *p* that measures the share of output that is ocean related. The sum of the ocean-shared industrial output is the ocean economy for the nation/region. This approach *disaggregates* the existing data. The other approach is to identify specific establishments where economic activity takes place as ocean related using some combination of industry and geographic location and calculating their output. This approach *aggregates* the existing data. Each approach has advantages and disadvantages, and both are currently in use. The two approaches are not mutually exclusive; they may be combined, but one of the approaches must be selected as the principal strategy. The United States uses both approaches, so it is a useful case study.

The Economics: National Ocean Watch (ENOW) data set[16] has been produced since 2002 and was the first ocean economy data set to provide consistent measurement across multiple years from the regional (county and state) level to the national level. The data are derived from the unemployment insurance system in the United States. Data from over nine million employment locations are collected every quarter, showing the number of people in employment each month and total wages paid. Using industrial and geographic criteria (discussed below) establishments in 30 coastal states are classified as being in an ocean-related industry or not. Value added is provided from a separate source and is assigned to each establishment using a disaggregation formula.[17] This process yields annual estimates of employment, establishments, wages paid, and value added at the regional level (county and state) and at the national level when calculated across all ocean-related establishments.

In 2020 the US government added a top-down approach to improve the coverage and accuracy of the national estimates.[18] This method begins with the product level data maintained by the US Bureau of Economic Analysis (the agency responsible for producing data about the US gross domestic product – GDP). Estimating GDP requires the use of input-output tables to track all the inter-relationships among industries, consumers, governments and foreign trade.[19] Most input-output tables are based on industry-based relationships, but this particular project used a product level table with 8,000 rows and columns. Using an expanded list of industries defined as ocean related based on the detailed product data, ocean relationship proportions called partials were identified.

Estimating partials is the central task of the top-down approach. A partial is simply a ratio (percentage) of the output of a product/industry that expresses the share of output that is ocean related. Partials must be estimated from a wide variety of sources and must often be approximated. Some US ocean-related industries, such as commercial fishing, are entirely marine and therefore have a partial of 1.0.

Ships and boats are obvious candidates for the ocean economy, but boats may be used on inland and marine waters. Boats for marine use may be built inland, and vice versa. Previously, in the United States inland versus marine use of boats and ships was determined using registration data to determine the location of the registered owners. This was imperfect but a high correlation between location of owner and area of use was expected.

Tourism and recreational activities in the coastal area are another key ocean industry; while the bottom-up approach uses the location of hotels, for example, location hides the distinction between travel for business and travel for leisure. The latter terminology defines ocean tourism, so surveys of travellers were used to distinguish between the different types of travellers.

The two approaches for estimating the ocean economy can be compared as follows:

Table 11.1 Two approaches for estimating the ocean economy

Top-down	*Bottom-up*
Higher precision in the ocean relationship	Lower precision in the ocean relationship
More data intensive to calculate partials	Only two data sets needed
National level data	Regional level to national level data

Industry

Other chapters in this volume are a good starting point for identifying the industries that are ocean related, but the selection of industries is more complicated. Most ocean economy accounts define ocean-related industries and sectors (aggregates of similar industries). A review of 27 different definitions of the ocean economy shows that the data can be organized into 10 sectors.[20] These are shown in Table 11.2.

Table 11.2 Ocean-related sectors

Marine transportation	*Shipping and maritime transport*
Search and navigation equipment	International cruise industry
Ports	Ports, transport and logistics
Transport services	Passenger water transport
Water-based transport of passengers and freight	Freight water transport
Shipping and port industries	Maritime transport

The marine transportation sector can be used to illustrate the variety of approaches to define the industries of the ocean economy. There are 12 different ways of identifying the industries in the marine transportation sector. Some of these labels probably cover the same activities, for example 'marine transportation' and 'maritime transport'. But others may have different meanings. Cruise ships are separated from water-borne transportation, and international cruise ships are separated from other types of cruises.

The variety of definitions of industries can be seen in Table 11.3 which is drawn from the study of 25 different ocean economy definitions noted above. For each of the 11 sectors, the number of industries included in that sector are shown, ranging from 1 to 12. The number of ocean economy definitions in which an industry is noted, along with the percentage of all ocean economy definitions in which an industry is mentioned. Some sectors, including government, living resources and marine transportation are included in one way or another in all of the ocean economy definitions, some in as few as one definition.

Two factors drive the choice of industries in the ocean economy. The first is the presence or absence of industries within a country; some industries, such as salt production and processing, are present and sometimes not. But the most important factor shaping the industries of the ocean economy are the industrial taxonomies used in the existing economic data. This too varies around the world. Most countries rely on the International Standard Industrial Classification (ISIC),[21] but there are variants of industrial taxonomies found in the People's Republic of China, the European Union, North America, Australia, New Zealand and elsewhere.

These taxonomies form the irreducible minimum of detail that can be used. For example, seafood processing can be divided between canned and frozen products. But if the industrial taxonomy has only a single classification for seafood processing, that is the smallest level of detail that can be observed. Increasing specificity in industrial taxonomy means increasing the level of detail in the measurement of the ocean economy. This is the reason why the US Ocean Economy Statistics series started with a product level input-output table with 8,000 rows and columns.

Table 11.3 Ocean-related industries

	Number of industries	*Number of mentions*	*% of mentions*
Marine construction	5	8	31%
Government	5	26	100%
Living resources	12	26	100%
Misc. marine products	1	5	19%
Marine services	12	22	85%
Research and education	2	15	58%
Ship and boat building and repair	6	23	88%
Tourism and recreation	9	24	92%
Marine transportation	12	26	100%
Utilities	6	19	73%
Technology	9	22	85%

Source: Charles S. Colgan (2020) *The U.S. Ocean Economy Statistics in International Context.* Monterey, CA: Center for the Blue Economy. www.centerfortheblueeconomy.org.

Geography

Geography is a key component of the ocean economy definition because location is itself a major factor in determining the extent to which an ocean relationship exists. A region must be defined within which a high degree of ocean relationship can be assumed to be likely and outside of which the relationship is assumed to be unlikely. The ocean relationship region must have an inland and a maritime boundary.

The obvious choice for defining the ocean-related region is the ‘coast’. But what is the coast? Of course, it includes the shorelines of the oceans, but how far inland does the definition extend? It might be an arbitrary distance of *x* km, but most economic data are maintained by politically defined jurisdictions such as provinces, which do not fit neatly with arbitrary distances. Nor do political jurisdictions necessarily follow natural features such as rivers and estuaries, and even if they do, where is the boundary between the ‘coastal’ and the ‘inland’ portion of a river? It may also be the case that certain inland bodies of waters have all the characteristics of the oceans from an economic perspective. Should such lakes be included in the definition of the ocean economy? The United States does in fact include the Great Lakes in its definition of the ocean economy, but the Great Lakes are not included in the Canadian definition even though the United States and Canada share the Great Lakes. In sub-Saharan Africa, Lake Victoria is an obvious candidate for inclusion in the ocean economy definition, although, like Lake Superior, it is situated thousands of kilometers upstream from the sea. Other large lakes such as Lake Tanganyika or Lake Malawi could also be considered.

The other boundary of the ocean economy zone is somewhat easier to identify. The rules governing the extent of national jurisdictions, or national maritime boundaries, have been the subject of Law of the Sea conventions since 1958. Since December 1982 coastal and island states have increasingly exercised, in a legal capacity, sovereign power over their adjacent maritime jurisdictional zones which include a territorial sea extending up to 12 nautical miles from the baseline of a coastal state (a country has complete jurisdiction over its territorial sea), and additionally a contiguous zone extending further from the outer edge of the territorial sea up to 24 nautical miles from the baseline and furthermore an exclusive economic zone (EEZ) which generally extends 200 nautical miles from the baseline; a state exercises more limited jurisdiction over its EEZ. There are some nuances to this boundary, such as the rules for jurisdiction on the continental shelf and some historical exceptions in certain countries, but the territorial sea, contiguous zone and EEZ are relatively clear maritime boundaries. Beyond the EEZ lie the high seas, which are governed by the 1982 Convention on the Law of the Sea. For many states in sub-Saharan Africa, the maritime area within these boundaries exceeds their land area (see Appendix).

Nevertheless, for legal purposes the extent of the maritime zone is not critical, since very few transactions take place 200 nautical miles from the shore (and thus 'within' the country) and 201 nautical miles from the shore (and therefore 'outside' the country). Fish caught or minerals extracted do not really enter the national economy until they are brought to shore. Fish caught beyond the EEZ and landed at a nation's ports are still counted in that nation's economy. There are some issues that will arise with jurisdictional boundaries in the extended accounts such as the question of who is responsible for sustainable fishing on the high seas. But the real difficulties lie in defining the inland boundaries, where there is a great deal of flexibility and no agreed rules. Each country must fit its ocean-related economic zone to its own unique geographic circumstances. The most important aspects of that decision are that it be clear and consistent.

Economic concepts

There are several options for measuring the ocean economy, but there are a few which are essential. The first is gross output (GO), which is the total output value of all final goods and services produced within the borders of a nation usually over the course of one year. This is the definition of GDP. GDP adjusts national output for internationally traded goods and services by subtracting imports from exports. This adjustment is necessary because GDP is usually measured at the point of final consumption by consumers, investors and the government. The GDP account also measures production by industry and the total production by industry should equal the total amount consumed (less what is imported).

The second essential measure is gross value added (GVA). GVA is gross output minus the costs of inputs and is equal to wages paid plus payments to owners adjusted for taxes paid. GVA is in many ways the preferred measure for measuring the ocean economy because using GO raises double counting issues. To take a simple example, the output of the fish harvesting sector could be the input to retail markets, in which case the retail price is the value of the fish. But if the harvested fish is sold to dealers who then sell to processors who then sell to wholesalers and then to retailers, the gross output of each step in this process cannot be added together because the output of one industry is the input to the next. This is why GVA is preferable; it measures only the *addition to value* at each stage. Calculated across all stages it reflects the entire value of the good or service. GVA is also preferable because it allows accounting for output in different places. The value of the fish harvested and landed in one province can be added to the value of the processing industry in another province and the value of the retail or restaurant in still another province.

Employment is technically not part of the national income accounts, but it is of great interest to policymakers and the public, so most countries report both output and employment. But employment is a tricky concept, with multiple possible definitions. Employment varies from month to month. In some industries it coincides with the seasons, and annual averages are usually reported. Employees may be engaged full time (in excess of a certain number of hours per week) or part time. They may be engaged for all of a year or only part of a year. Each country's statistical systems will have its own definitions of how employment is measured and for the most part this will determine the way employment is expressed in the ocean economy.

Wages are often included with employment, and wages do reflect to some extent the shares of part-time and part-year employment. Wages are a category of labour income, which is a broader term that reflects total compensation including benefits such as the employer's share of pension contributions. Wages are very important if ocean accounts are to be used in assessing equity outcomes, as will be discussed below.

The number of reporting establishments is another useful indicator. Changes in the number of establishments in an industry can be interpreted as a proxy for investment. Establishment-level measurement is the foundation for the bottom-up ocean economy measurement so this data should be incorporated in that process. In contrast, establishments are generally not included in top-down estimations and must be added from other sources if so desired.

Tax receipts are also of interest to policymakers. Nations with value-added tax systems have an advantage in that VAT systems can be used to measure ocean economic activity directly, converting from VAT receipts to GVA so long as the tax records contain industry and geographic classification data. This approach measures both contributions to national GVA and revenues to the government. The example given below showing the ocean economy estimates for Tanzania illustrate this approach.

Final and intermediate goods

The discussion in the previous session about the importance of avoiding double counting of output through the use of GVA raises an important issue that has implications for the definition of the industries in the ocean economy. Virtually all the definitions of the ocean economy include the output of the ship and boat building industries, fish harvesting and marine transportation. Referring back to Table 11.1, many definitions include industries within a marine service sector. But these are a mixture of intermediate products (sold as inputs for the production of other products) and final products (sold only to final users). Boats may be built and sold to the fishing industry (an intermediate good) or to consumers for recreational purposes (a final good). Fish may be either intermediate or final goods. The use of GVA allows for the correct measurement of the value of production among the industries, but important relationships are obscured, and descriptions of the ocean economy need to clarify the connections among the industries when boats, fish and restaurants are all included in the ocean economy.

Past and future

A final consideration in constructing ocean economy accounts is how far back one should go to create estimates and how to develop a process that can be continued into the future. A major problem that has historically plagued efforts to measure the ocean economy is that studies have been done for one year or two years and then discontinued, making it impossible to look back and detect trends or to monitor future changes that will be influenced by policy choices. This is important because, while the size of the ocean economy is definitely of interest, it is the changes in that economy that are particularly important for informing policy and for tracking the consequences of future policies. Short-term studies are also often undertaken using methods or data that may be difficult or impossible to replicate. A critical element of the design of ocean economy accounts is to use data so that a history can be constructed back to some base year and that will be available in the future for the continuous production of ocean economy data.

Notes on creating ocean economy accounts

There are three general notes about creating ocean economy accounts. The first is the need to build confidentiality protections into the data. All government data systems are built to one degree or another on voluntary reporting of information from businesses and other organizations. Even tax systems mostly rely on voluntary compliance because the costs of enforcement are high. Businesses provide information that could affect their sales and profits if their competitors acquire it. For this reason, government data systems employ tests to reduce the possibility that the value of any one company or organization could be revealed.

These tests vary across jurisdictions, but generally comprise a rule that data will not be released below some minimum number of reporting entities (if there are fewer than four reporting units in an industry, for example). A second rule prevents release if one entity dominates the data by making up, say, 80% of the combined total value added, employment or whatever data concept is being reported. There are many variations to these rules and their application to the ocean economy data will depend on the rules in place within the agencies supplying data. In general, the finer the geographic and industrial detail that is desired, the more data will need to be suppressed to protect confidentiality. Conversely, the larger the geography and the more aggregate the industries that are reported, the more data can be shown.[22]

A second note is the creation of satellite accounts, i.e. accounts that are completely consistent with a nation's national income and product accounts. The methods described here create ocean economy accounts that are highly useful and can be compared with the data for other industries. But some of the features of full national income accounts, particularly the measurement of prices and quantities, are not found in the ocean economy accounts. Separating prices and quantities is an important part of national income accounting because a change in output or value added from one period to another may consist of a change in the quantity sold or simply a change in price. The first represents real growth, the second is inflation. The long-term objective of ocean account creation is that it should be fully integrated within the national income accounts; this may take considerable time, so beginning with one of the two approaches described above can provide very useful information in less time and at a lower cost.

A final note is that the process of estimating ocean accounts continues; innovative approaches are being developed and tested in a number of places and by organizations other than national governments or academic researchers. Two examples are a report on the ocean economy in the Indian Ocean prepared for the World Wildlife Fund[23] and a report by the Organisation for Economic Co-operation and Development.[24] These studies are concerned with multinational estimates of the ocean economy, but they start with national level estimates and provide additional methodological approaches for specific industries.

An example of constructing an ocean account: Tanzania

An example of the process of constructing an ocean account is provided in Table 11.4, which provides a very simple disaggregation estimate of an ocean economy for Tanzania. This table uses two criteria for defining the ocean economy: industry (as defined by the ISIC) and geography (economic activity in coastal provinces). The table shows whether data from coastal provinces or national sources are used. There are also two sources for the data, published value added data from the Bureau of National Statistics and the VAT revenues converted to value added using the inverse of the tax rate.

The choice of industries for measurement is based on similar preliminary ocean economy studies carried out for Bangladesh and several countries in the Caribbean.[25] The data indicate that, consistent with many other countries, industries in tourism and recreation, which are measured here only in the coastal provinces, make up the largest share of the ocean economy. The total ocean economy is shown in both current prices and adjusted for inflation using the World Bank inflation data. The ocean economy in Tanzania grew by 84% between 2015 and 2019 in current prices and by 58% in inflation-adjusted prices, and this an *underestimation* of that growth since there are some sectors for which 2019 data are not available. It suggests that the ocean economy has been an important part of growth in the Tanzanian economy.

This analysis also shows many of the issues that can be identified in a first draft compilation. Some sectors, such as extraction of natural gas are shown as not being present. This may well be the case, but it is an example of an industry that could be double-checked in the data. The same is true of marine fishing. Tanzania has tracked inland fisheries as well as marine fisheries. The inland fisheries may come from Lake Victoria, which raises the question of whether Lake Victoria should be in the ocean economy or the blue economy. It should be noted that the VAT revenues may distort the estimation of VA because of unknown special provisions of the definition of the VAT base in some industries. But the VAT data have the ability to make geographic distinctions critical to the ocean economy.

Table 11.4 is not an official estimate of the Tanzanian ocean economy but it is evidence of how existing data series can be organized to provide at least a preliminary estimate of the ocean-related part of national income and to identify the next steps for improvement.

Extending the ocean economy to the blue economy

Measurement of the ocean economy provides an important but still partial view of the relationship between the ocean and society. Changes in the level of employment or output in fisheries may be interpreted as a decline in fish stocks, a shift in consumer tastes, or a decline in prices. Planning for the future of a sustainable fishing industry depends on knowing the proportion of these effects on stocks of fish, on the habitats in which the stocks mature, and on the related and supporting industries for the fishing industry. This larger vision of the economy comes closest to the blue economy. The statistical framework to provide this larger vision is largely in place because of the developments in economic accounting over the past two decades. That framework is being specifically adapted to ocean economies and ecosystems in a new effort by the UN. Those nations wishing to invest in the measurement of the blue economy will find the foundation already well prepared.

The key to the expanded vision is two accounting frameworks that expand the System of National Accounts-based ocean economy measures discussed above. These are the System of Environmental-Economic Accounts-Central

Table 11.4 Estimated value added in the Tanzanian ocean economy (Tanzanian shillings, millions); figures in italics estimated from VAT

	ISIC	*Coastal*	*National*	*2015*	*2016*	*2017*	*2018*	*2019*
Marine fishing	311	X		1,871,672	2,056,008	2,245,558	Missing	Missing
Extraction of natural gas	620		X					
Extraction of salt	893		X	5,071	4,806			
Support activities for petroleum and natural gas extraction	910		X					
Support activities for other mining and quarrying	990		X	3,659,599	4,975,991	6,573,059	Missing	Missing
Processing and preserving of fish, crustaceans and molluscs	1020	X		*229,534*	*739,248*	*40,784*	*174,084*	*134,502*
Manufacture of vegetable and animal oils and fats	1040	X		*3,220,798*	*1,376,832*		Missing	Missing
Manufacture of prepared meals and dishes	1075		X	*171,153*	*151,603*	*334,144*	*1,582,050*	*1,230,616*
Building of ships and floating structures	3011	X		*492,795*	*998,363*	*1,354,335*	*2,014,989*	*2,198,092*
Building of pleasure and sporting boats	3012	X						
Repair of transport equipment, except motor vehicles	3315		X	*49,531*	*143,986*	*173,616*	*1,116,667*	*3,748,468*
Water collection, treatment and supply	3600		X	392,557	422,698	519,909	Missing	Missing
Sewerage	3700		X					
Materials recovery	3830			*17,096,448*	*22,256,209*	*19,929,409*	*20,026,711*	*19,005,611*

	ISIC	*Coastal*	*National*	*2015*	*2016*	*2017*	*2018*	*2019*
Sea and coastal passenger water transport/transport and storage	5011	X		3,864,481	4,473,876	8,381,276	Missing	Missing
Sea and coastal freight water transport	5012	X		*422,404*	*142,028*	*817,487*	*212,783*	*619,333*
Warehousing and storage	5210	X						
Service activities incidental to water transportation	5222		X	*6,317,853*	*2,311,900*	*2,346,259*	*6,147,576*	*2,569,232*
Cargo handling	5224	X						
Short-term accommodation activities	5510	X		*30,651,508*	*41,118,529*	*46,617,686*	*58,521,905*	*56,345,494*
Camping grounds, recreational vehicle parks and trailer parks	5520	X						
Other accommodation	5590	X		1,421,916	1,523,035	1,602,543	Missing	Missing
Restaurants and mobile food service activities	5610	X		*28,429,001*	*38,262,837*	*39,762,483*	*41,706,604*	*45,342,142*
Beverage serving activities (manufacture)	5630	X		*241,607*	*4,284,770*	*2,789,292*	*846,420*	*50,370,625*
Other amusement and recreation activities n.e.c.	9329	X			160			447
Total at current prices				98,537,929	125,242,879	133,487,841	132,349,788	181,564,563
Total at inflation adjusted prices (2011 base)				74,611,415	90,971,029	92,121,323	88,116,800	117,830,755

Source: Authors' calculations

Framework (SEEA-CF)[26] and the System of Environmental-Economic Accounts Experimental Ecosystem Service Accounts (SEEA-ES).[27] These inter-related accounts can be visualized in Figure 11.1.

The first step outwards from the standard national income accounts, which is termed the central framework of the environmental accounts, adds measures of the value of natural capital used as raw material inputs to the production goods and services and adjusts the values in the national income effects for the negative effects of pollution. The process can be briefly described as follows.

Natural capital raw materials

There are two principal natural capital raw materials in the ocean economy: living resources and mineral resources. To continue the fishing example used above, the stock (population) of fish is the fundamental raw material to the fishing industry and all the value added from the harvesting stage onwards to the final consumer. That stock has a value, just as real estate or buildings have a value. That value is determined by the size of the stock (biomass) and the price at which harvested fish can be sold. These can be forecast and the present value of future output (usually calculated at the harvesting stage) equals the value of the stock at any one time. That value changes each year. Fish are born into the population (recruited), thereby increasing the stock. This is offset by fish dying of natural causes or because they are caught.

The Environmental Account calculates the change in the value of the fish stock each year, which may increase if recruitment exceeds mortality, or it may decrease if the opposite occurs. The change in the stock (called a flow of value) is analogous to the change in the stock of capital in the fishing industry measured by additions and depreciation of boats, buildings, etc. The result of the combination of the national income and environmental account perspective is a measure of the annual changes in the value of the fishing industry *and* the fish stocks upon which it relies.

There is one more adjustment that the Central Framework accounts make: the stock of fish could be stable or even growing, but pollutants entering the water make the fish unsuitable for human consumption either entirely or above some concentration of pollution. This could be the case with large quantities of untreated or poorly treated sewage or agricultural run-off, for example. The other part of the Central Framework adds to the value of the economy the expenditures on reducing pollution by both the public and private sectors. Expenditures by the public sector on pollution control are embedded in the government sector as a whole, while expenditures by the private sector are normally accounted as inputs to the production of goods and services (and outputs for industries supplying environmental protection) and never appear in the country's GDP. The Central Framework calls for extracting these expenditures as showing them as separate tables indicating the level of commitment to the tasks of reducing pollution.

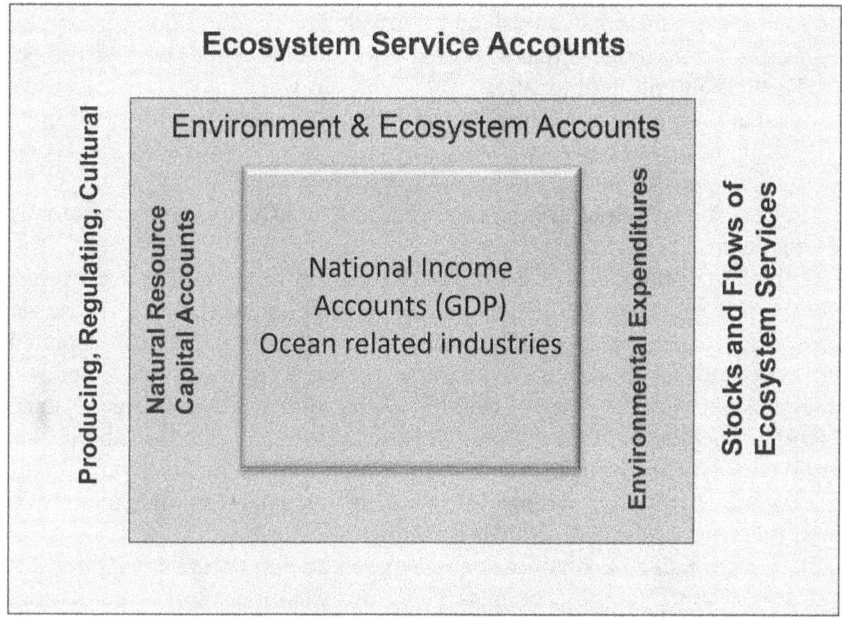

Figure 11.1 Accounting framework
Source: Authors' illustration.

The Ecosystem Accounts push the view of the economy one step further back to look at the ecological characteristics that underpin economic activity and which add value in ways not accounted for at all in the core national income accounts. The starting point is the concept of ecosystem services, which are simply the effects that ecosystems have on humans. The array of such effects is large, complex and until recently generally unappreciated because for the most part there were no easy measures of economic value. An example is wetlands, which traditionally have been considered as wasteland without value. Today it is understood that wetlands provide a large and diverse array of valuable services to people in all parts of the world.[28]

The array of ecosystem services has been organized into a taxonomy called the Common International Classification of Ecosystem Services (CICES).[29] Version 5.1 of CICES, which was released in 2018, divides the array of ecosystem services into three major and two subcategories. The major categories are as follows:

- Provisioning: all the nutritional, non-nutritional material and energetic outputs from living systems. Water and non-water related services are distinguished from one another.

- Cultural: all the non-material, and normally non-rival and non-consumptive outputs of ecosystems that affect the physical and mental states of people.
- Regulation and maintenance: all the ways in which living organisms can mediate or moderate the ambient environment that affects human health, safety or comfort.[30]

The two subcategories are the biological and non-biological characteristics of ecosystems.

With the development of an organized way of thinking about ecosystem services has come more intense focus on the economic values of those services, often expressed as concerns about natural capital. Natural capital is the same concept as discussed above in the value of fish stocks, but with ecosystem services the concept extends to the service outputs of ecosystems. Examples include the value of wetlands as nursery habitats for fisheries and birds, nutrient absorption and water quality improvements from aquatic grasses, sequestration of carbon to reduce carbon emissions, or reductions in flood damages behind sand dunes or mangrove forests.

There is, however, a significant challenge in adding ecosystem service flow and stock values to the accounts. There are well-understood and standardized ways of measuring the economic values in the national income and Central Framework accounts, but this is not the same with ecosystem services. These values must be measured through a complex set of economic methods that must capture values not easily observable in transactions or not observable at all. Economists refer to many of the values of ecosystem services as 'non-market' values.

This complexity of methods is exacerbated by the fact that all the estimates used for ecosystem service valuations must be done in single studies of specific places. And these studies are almost always time-consuming and costly to undertake to the point that it is impractical to routinely estimate such values and in many instances studies from comparable locations and circumstances. Large databases of such values have been assembled. The distribution of such studies tends to be skewed towards North America and Europe,[31] but there are examples of relevant studies in sub-Saharan Africa such as ecosystem services of seagrasses and mangroves in Kenya.[32] There have also been concerted efforts to compile ecosystem value data in developing countries. The World Bank has been particularly active in helping countries to recognize and make decisions to maximize natural capital.[33] The European Union and other organizations have prepared data and procedures for estimating and using natural capital vales under the project name the Economics of Ecosystems and Biodiversity (TEEB).[34]

The three layers of economic information provided by the national income account, the Central Framework Environmental Account and the Ecosystem Environmental Account provide a broad set of measures about the performance of the blue economy both as a contributor to economic growth and sustainability. These different measures can all be tracked separately. A recent report for the High Level Panel for a Sustainable Ocean Economy[35] suggests that the various measures in ocean accounts can be arranged in statistical 'dashboards', or online in easy to understand displays of the data. This report also discusses how the information in the accounts can be combined into simple measures of sustainability for the blue economy as a whole – an idea which has been explored in other studies.[36]

The key to this approach is to understand the role of capital, defined here as the stock of assets that are combined to produce goods and services. Traditionally, these are land, labour and capital. To this list is added natural capital. National income accounts measure changes to those stocks through investment and depreciation. An important measure in the national income accounts, though one rarely cited, is the net domestic product (NDP), which is the GDP minus an allowance for depreciation of buildings, equipment, etc. The GDP measure of capital investment simply considers the total amount spent on buildings, land, equipment, etc. But some amount of this gross investment just pays to offset depreciation. If GDP investment equals depreciation, an economy has added nothing to its capital to support future growth. If investment is less than depreciation, future economic activity will decline. Only if the stock of capital increases faster than depreciation will the economy grow in the future.

This principle extends to natural capital and the importance of the environmental account becomes apparent. If future sustainable growth from the national income perspective depends on the net rate of capital growth in physical capital, so also does it depend on net growth in natural capital. This includes both the natural capital measured in the Central Framework account, such as the value of fish stocks, and the natural capital of the ecosystem services in the Ecosystem Accounts. In these latter cases, the availability of the data from fully implemented Environmental Accounts (both Central Framework and Ecosystem) allows calculation of net environmental capital and net ecosystems capital (together making up net natural capital). When net environmental and ecosystems capital are growing, sustainable growth in the overall economy is possible. When net environmental and ecosystems capital is declining, as is too often the case, sustainable growth and sustainable core national income economies become impossible.

Again, this critical information comes at a cost in terms of time and resources. It is difficult to implement because it is time-consuming and costly to do so. Whether a top-down or bottom-up approach is taken to estimate the core ocean account, it takes time to acquire and organize the necessary data. Extending to the environmental accounts requires collection of a great deal of data that must be collected in systems that for the most part do not exist or do not exist in an organized way.

The current income and environmental accounts are also incomplete. It may have been noted that of the three principal factors of production (land, labour and capital) only two (land and physical capital such as equipment) are measured in the national income accounts. Labour is simply an input on the production side and compensation is tracked on the income side of the National Income and Products Accounts as income to households. This view of labour has undergone a radical transformation over the past few decades with a shift to thinking of labour as human capital. In this new view, human capital is the sum of the knowledge, skills and abilities that labour contributes to production, and not simply the sum of hours worked. Human capital, like natural or physical capital, can be increased or decreased. In general, increasing human capital through increasing education and skills is as much a prerequisite of sustainable growth as is increasing physical and natural capital.[37]

Human capital is not, however, included in the UN accounting systems. The measurement of physical capital is widely understood and more or less consistently applied. With natural capital there is general understanding of theory and methods of measurement, albeit that theory and methods are applied in wildly uneven ways. But measurement of human capital has neither agreed upon theory or methods. The addition of human capital to the economic accounts and to the blue economy accounts is a major challenge for the coming decade. The World Bank has developed a Human Capital Index (HCI) which it calculates for all countries. The HCI is measured at the national level, but it suggests at least one approach to measuring human capital that could be adapted to the blue economy as a starting point.[38]

The other major aspect of sustainability that the economic accounts cannot address on their own is equity. Starting with the Brundtland Commission declaration on sustainability emerging from the original commission on sustainability in 1987, equitable outcomes have been foundational to definitions of sustainability.[39] Nine of the 17 Sustainable Development Goals announced by the UN in 2015 specifically address equity issues, and all of the goals touch on equity.

The appropriate definition of equity for the blue economy is a contested subject; it is not clear that there will ever be agreement on a universal definition. Equity will have to be defined in the specific contexts of local history, culture and economic circumstances. As such, the process of measuring the blue economy can provide information about progress towards whatever equitable goals are set. Ensuring the inclusion of data relating to labour, details on full- and part-time employment and on wages will be a useful measurement whatever the equity goals and should be included in the ocean economy measurements. Similarly, data for the blue economy's functioning at the subnational regional level (provinces, urban/rural areas, etc.) can be useful for equity considerations. The important point is that the ability to address equity considerations need to be intentionally built into the process of constructing the national income and environmental accounts from the outset.

Conclusion

Measuring the blue economy is often overlooked in the early announcements touting the importance of the oceans and coasts to national economic futures. But measurement is the key to accomplishing whatever goals are set. Indeed, it is critical for defining those goals in the first place. There are clear advantages to developing blue economy data today compared with the past. There is substantial experience and expertise in organizations around the world. The UN is making an important effort to explicitly adapt its economic and environmental accounts to the specific needs of ocean accounting. This effort is still in its early stages, but the UN has developed a technical guidance document that provides detailed information about how to construct ocean accounts, environmental accounts and ecosystem service accounts. A secretariat provides a central contact point for efforts and expertise in the growing global ocean accounting community.[40]

Nonetheless, the prospect of collecting, analysing and maintaining all the data needed to provide a full measurement of the blue economy is daunting. It is as if one were in Banjul in The Gambia or in Mogadishu in Somalia and one was told to sail a 6 m pirogue (on the west coast of Africa) or a dhow (on the east coast) to the Cape of Good Hope. Looked at from the map of Africa, the prospect of sailing across that vast space in a small craft appears to suggest 'don't go!'. But if one remembers the ultimate destination and simply sails along the shore from day to day, which is the traditional approach to navigation in Africa, then it is certainly possible to get to the Cape.

So, the advice here is to just start out. The data needed to set up the core ocean economy accounts are already somewhere in the national statistical systems. It will take some effort to organize, but even the crudest measurement of the ocean economy is a start that will point to the next steps. Gaps will be apparent in the data that are now available and strategies for expanding and refining that data will become clear. Each step ahead points to the ones needed after that and both short- and long-term plans can be set. Budgets will always be a problem; there will never be enough money to do everything needed, at least to the schedule desired. But, like weather on a voyage, this is to be expected and alterations can be made as needed. If there are budget limitations, there is also the growing pool of experience and expertise to help make the most efficient use of the resources that are available.

Ultimately, the real test of commitment to the principles of a sustainable use of ocean resources is not the announcements of policy or even the expenditure of funds. It is the willingness to develop the systems that measure progress, including the ability to measure failures. A country that cannot say what is changing in its use of the ocean and its resources is unlikely to make the difference in the oceans or its economy it says it wants.

Appendix

Table 11.5 Coastal/Oceans Area Measurements and Relationships

Nation	*EEZ (sq km)*	*Continental shelf (sq km)*	*Land area (sq km)*	*Ocean relationship*	*Population (2019)*
Angola	518,430	48,090	1,246,700	CS	32,866,272
Benin	33,220	2,720	112,760	CS-ZLS	12,123,299
Cameroon	16,550	11,420	472,710	CS	26,545,860
Cabo Verde	800,560	5,600	4,030	IS	555,987
Comoros	163,750	1,520	1,860	IS	869,000
Congo Republic	31,020	7,980	342,000	CS-GDS	5,518,087
Democratic Republic of the Congo	1,600	1,590	2,267,050	CS	89,561,403
Djibouti	7,460	3,190	23,200	CS-ZLS	998,000
Equatorial Guinea	303,510	7,820	28,050	CS	1,420,985
Gabon	202,790	35,020	257,760	CS	2,224,724
Ghana	235,350	22,500	227,540	CS	31,072,940
Guinea	59,420	44,750	245,720	CS	13,132,795
Guinea-Bissau	123,720	39,340	28,120	CS-ZLS	198,600
Côte d'Ivoire	176,250	10,170	318,000	CS	26,545,864
Kenya	116,940	11,070	569,140	CS	53,771,296
Liberia	249,730	17,710	96,320	CS	5,057,681
Madagascar	1,225,260	101,500	581,800	IS	27,691,018
Mauritius	1,285,000	29,060	2,030	IS	1,271,768
Mozambique	578,990	94,210	786,380	CS	31,235,435
Namibia	564.748	86,698	823,290	CS	2,540,905
Nigeria	217,310	42,290	910,770	CS	206,139,589
São Tomé and Príncipe	131,400	1,900	960	IS	219,519
Senegal	158,660	23,090	192,530	CS	16,743,927
Seychelles	1,336,560	39,060	460	IS	93,847
Sierra Leone	215,610	28,620	72,180	CS	7,976,982
Somalia	825,050	55,900	627,340	CS	15,893,222
South Africa	1,535,540	156,337	1,213,090	CS	59,308,690
Tanzania	241,890	25,610	885,800	CS	59,734,218
The Gambia	23,110	5,580	10,120	CS-GDS	2,416,688
Togo	12,040	1,260	54,390	CS-ZLS	8,278,724

Key: CS = Coastal State; IS = Island State; GDS = Geographically Disadvantaged State; ZLS = Zone Locked State.
Note: The latter two categories are special categories in which the rights to maritime space are unclear under the 1982 Convention on the Law of the Sea. Cf. Lewis M. Alexander and Robert D. Hodgson, The Role of the Geographically Disadvantaged States in the Law of the Sea, 13 SAN DIEGO L. REV. 558 (1976).

Notes

1 United Nations Conference on Sustainable Development (Rio+20), paras 2, 56.
2 World Bank and United Nations Department of Economic and Social Affairs. 2017. *The Potential of the Blue Economy*; UNEP, FAO, IMO, UNDP, IUCN, GRID-Arendal (2012) *Green Economy in a Blue World*. www.unep.org/greeneconomy.
3 Patil, Virdin, Colgan, Hussain and Failler, Vegh. 2018. *Toward a Blue Economy: A Pathway for Bangladesh's Sustainable Growth*. May, p. 109. http://documents.worldbank.org/curated/en/857451527590649905/Toward-a-blue-economy-a-pathway-for-.
4 Patil, P. G., Virdin, J., Diez, S. M., Roberts, J. and Singh, A. 2016. *Toward A Blue Economy: A Promise for Sustainable Growth in the Caribbean; An Overview*. Washington, DC: World Bank. http://documents.worldbank.org/curated/en/965641473449861013/pdf/AUS16344-REVISED-v1-BlueEconomy-FullReport-Oct3.pdf.
5 United Nations Development Programme. 2018. *Leveraging the Blue Economy for Inclusive and Sustainable Growth in Kenya*. www.icriforum.org/meeting/east-asian-seas-eas-congress-2012.
6 United Nations Economic Commission for Africa. 2016. *Africa's Blue Economy: A Policy Handbook*. Addis Ababa: UNECA.
7 African Union. 2012. *2050 Africa's Integrated Maritime Strategy*. Addis Ababa, AU.
8 Inter-African Bureau for Animal Resources (AU-IBAR). 2019. *Africa's Blue Economy Strategy*.
9 Nathan Associates. 1974. *Gross Product Originating from Ocean-Related Activities*. Washington, DC.
10 Colgan, Charles. 2004. 'Employment and Wages in the Ocean Economy of the United States'. *Monthly Labor Review*. November; Colgan, Charles. 2007. *A Guide to the Measurement of the Market Data for the Ocean and Coastal Economies*. Monterey, CA. www.oceaneconomics.org.
11 Gardner Pinfold. 2009. *Economic Impact of Marine Related Activities in Canada*. Ottawa.
12 Allen Consulting Group. 2004. *The Economic Contribution of Australia's Marine Industries 1995–95 to 2002–03*. https://parksaustralia.gov.au/marine/management/resources/scientific-publications/economic-contribution-australias-marine-industries-1995-96-2002-03/.
13 Statistics New Zealand. 2016. *New Zealand's Marine Economy: 2007–2013. SEEA Account*. Wellington.
14 In some countries large freshwater bodies are included in the definition of the ocean economy.
15 United Nations Department of Economic and Social Affairs. *National Accounts: A Practical Introduction. Studies in Methods Handbook of National Accounting*. New York: DESA.
16 Colgan, Charles S. 2013. 'The Ocean Economy of the United States: Measurement, Distribution, & Trends'. *Ocean & Coastal Management* 71(1): 1–10. doi:10.1016/j.ocecoaman.2012.08.018.
17 The disaggregation equation for an individual establishing in industry i is $V_e^i = V_s^i\left(\frac{w_e^i}{w_s^i}\right)$ where V_e^i = value added in industry i in establishment e; V_s^i = Value added in state s for industry i; w_e^i = wages in establishment e in state s; and w_e^i = wages in establishment e in industry i. The total industry output is then $\sum_1^n\left(V_e^i\right)$.
18 Nicolls, William, Connor Franks, Teresa Gilmore, Rachel Goulder, Luke Mendelsohn, Edward Morgan, Jeffery Adkins and Charles Colgan. 2020. *Defining and*

Measuring the U.S. Ocean Economy. Washington, DC: US Department of Commerce, Bureau of Economic Analysis.

19 UN Department of Economic and Social Affairs. 1999. *UN Handbook of Input-Output Table Compilation and Analysis. Studies in Methods Handbook of National Accounting Series F No 74*. New York: DESA; Miller, R. and P. Blair. 2009. *Input-Output Analysis, Foundations and Extensions.* Cambridge, MA: Cambridge University Press.

20 Colgan, Charles S., and Jeremy Ginsberg. 2020. *The U.S. Prototype Ocean Economy Statistics in International Context*. Charleston, SC: Middlebury Institute of International Studies, Center for the Blue Economy.

21 United Nations Statistics Division. 2008. 'International Standard Industrial Classification of All Economic Activities Rev.4'. http://unstats.un.org/unsd/cr/registry/isic-4.asp.

22 A brief non-technical explanation of the US confidentiality rules can be found in Colgan, 'The Ocean Economy of the United States'.

23 Obdura, D. 2017. *Reviving the Western Indian Ocean Economy: Actions for a Sustainable Future.* Gland: World Wide Fund for Nature. doi:10.2307/20304954.

24 Organisation for Economic Co-operation and Development. 2016. *The Ocean Economy in 2030*. Paris: OECD. doi:10.1787/9789264251724-en.

25 See notes 3 and 4.

26 United Nations, European Union, Food and Agriculture Organization of the United Nations, International Monetary Fund, Organisation for Economic Co-operation and Development, and World Bank. 2014. *System of Environmental-Economic Accounting: A Central Framework*.

27 United Nations Statistics Division. 2014. *System of Environmental-Economic Accounting: Experimental Ecosystem Accounting*. doi:10.5089/9789211615630.069. This experimental system is currently being revised as a final document; the full Ecosystem Account was released in early 2021. For more information see www.seea.un.org.

28 Millennium Economic Assessment. 2005. *Ecosystems and Human Well-Being: Current State and Trends, Volume 1*. Washington, DC: Island Press.

29 www.cices.edu.

30 Haines-Young, Roy, and Marion Potschin. 2018. *CICES V5. 1. Guidance on the Application of the Revised Structure.* www.cices.eu.

31 Environment Canada has one of the best online searchable databases of ecosystems service values. See www.evri.ca.

32 Nyunja, J., M. Ntiba, J. Onyari, K. Mavuti, K. Soetaert, and S. Bouillon. 2009. 'Carbon Sources Supporting a Diverse Fish Community in a Tropical Coastal Ecosystem (Gazi Bay, Kenya)'. *Estuarine, Coastal and Shelf Science* 83(3): 333–41. doi:10.1016/j.ecss.2009.01.009; Kairo, James G., Caroline Wanjiru and Jacob Ochiewo. 2009. 'Net Pay: Economic Analysis of a Replanted Mangrove Plantation in Kenya'. Journal of Sustainable Forestry 28 (3–5): 395–414. doi:10.1080/10549810902791523.

33 World Bank. 2016. 'Managing Coasts with Natural Solutions', January: 167. www.wavespartnership.org/en/knowledge-center/managing-coasts-natural-solutions; WAVES Partnership. 2016. 'Wealth Accounting and the Valuation of Ecosystem Services', www.wavespartnership.org/en.

34 Van der Ploeg, S., Dolf De Groot and Yafei Wang. 2010. 'The TEEB Valuation Database: Overview of Structure, Data and Results'. *Foundation for Sustainable Development, Wageningen, the Netherlands*, December: 248. Sukhdev, Pavan, Heidi Wittmer, Christoph Schröter-Schlaack, Carsten Nesshöver, Joshua Bishop, Patrick ten Brink, Haripriya Gundimeda, Pushpam Kumar and Ben Simmons. 2010. *The Economics of Ecosystems and Biodiversity: Mainstreaming the Economics of Nature: A Synthesis of the Approach, Conclusions and Recommendations of*

TEEB. Environment. www.iges.or.jp/jp/news/topic/pdf/1103teeb/teeb_synthesis_j.pdf.

35 Fenichel, Eli P, Ben Milligan, Ina Porras, Ethan T Addicott, Ragnar Árnasson, Michael Bordt, Samy Djavidnia *et al*. 2020. 'National Accounting for the Ocean and Ocean Economy', 1–48, www.oceanpanel.

36 Colgan, Charles S. 2016. 'Measurement of the Ocean Economy from National Income Accounts to the Sustainable Blue Economy'. *Journal of Ocean & Coastal Economics* 2(2).

37 Blaug, Mark. 1976. 'The Empirical Status of Human Capital Theory: A Slightly Jaundiced Survey'. *Journal of Economic Literature* 14(3): 827–55. www.jstor.org/stable/2722630 (accessed 1 November 2020).

38 www.worldbank.org/en/publication/human-capital.

39 United Nations World Commission on Environment and Development. 1987. *Our Common Future*. New York: United Nations.

40 See www.oceanacounts.org.

12 International ocean governance as a necessary backdrop to developing Africa's blue economy

Nick J. Hardman-Mountford, PhD, Jeff A. Ardron, MSc, Rosemarie Cadogan, MSc, Chilenye Nwapi, PhD and Alison Swaddling, MA

Introduction

Since its promotion at the 2012 United Nations (UN) Conference on Sustainable Development (UNCSD, often referred to as Rio+20),[1] the sustainable blue economy concept has been gaining traction across Africa, championed by several national governments and regional intergovernmental organizations.

Currently, the blue economy in Africa generates US $296 billion per annum, and it is predicted that by 2030 this figure will have increased by 37% and that it will have almost doubled by 2063. These revenues support 49 million jobs, and by 2063 this number is expected to increase by approximately 60% to nearly 80 million jobs (AU-IBAR, 2019). The major sectors underpinning this projected growth are fisheries, aquaculture, international shipping, tourism and offshore oil and gas. Expected new growth areas include offshore renewable energy, regional shipping, seabed mineral extraction, 'blue carbon' and other natural capital investments, education, research and technology.

Although not necessarily sustainable in their current formulations, these sectors are seen as 'blue' and integral to attaining the pan-African vision of the African Union (AU)'s 'Agenda 2063: The Africa We Want', a 50-year 'blueprint and masterplan for transforming Africa into the global powerhouse of the future' (AU-IBAR, 2019). Specifically, goal 6 of Agenda 2063 focuses on the 'blue/ocean economy for accelerated economic growth' (AU, 2015). In contrast, UN Sustainable Development Goal (SDG) 14: 'life below water' aims to 'conserve and sustainably use the oceans, seas and marine resources for sustainable development' (emphasis added).[2] The different emphases of these goals demonstrate one of the key challenges of Africa's blue economy vision – how to manage ocean resources sustainably while also exploiting them to accelerate economic growth.

To date, the multitude of international commitments for sustainable development and ocean management have failed to be implemented sufficiently to achieve the prerequisite regional and national governance frameworks, with ocean financing falling far short of what would be required to meet SDG 14 (Johansen and Vestvik, 2020) and few of the 2020 UN biodiversity and ocean sustainability targets being reached.[3] Additionally, there are governance challenges across sub-Saharan Africa in implementing blue economy initiatives, whereby existing local economies are sometimes disregarded and placed in jeopardy, thereby undermining the intent of these top-down 'blue' initiatives (Okafor-Yarwood *et al.*, 2020). Thus, ensuring durable blue economic development in accordance with international and regional agreements in Africa presents not only the familiar conundrum of 'sustainable development' (McCloskey, 1998) but more fundamentally the question of its governance and resourcing.

What is the blue economy? An African view

Key to understanding the potential and the challenges of the blue economy is a clear understanding of what it means. For some, the 'blue' label refers solely to the aquatic focus of a country's activities, whereas for others it acts as an aquatic parallel to the 'green' moniker of the green economy, denoting a focus on environmental sustainability. Furthermore, the scope of aquatic activities can range from inclusion of seas, lakes and rivers (the AU perspective), to a purely marine focus (such as that taken by Agenda 2063). The sustainability perspective has certainly been integral to the development of the blue and green economy concepts within the international community, both being popularized through the UNCSD. The emerging consensus is that the blue economy refers to economic activities arising from sustainable use of the ocean, and in some cases other water bodies, with full consideration given to ecological and social values (e.g. UNCTAD, 2014; UNEP, 2015; AU, 2015; UNECA, 2016). Perhaps the best definition of a successful blue economy in the African context is one that 'emphasizes the interconnectedness of economic development, local community inclusion and environmental sustainability without prioritizing one aspect over the other' (Okafor-Yarwood *et al.*, 2020).

Most advocates of the blue economy include within its scope the major sectors of offshore renewable energy, fisheries and aquaculture, ports and shipping, and tourism. Some also include offshore oil and gas and deep-sea mining. Currently, the sustainability of these various sectors varies. How far a strictly sectoral approach to governance will deliver on the global goals of the 2030 sustainability agenda remains to be seen. However, an integrated approach is often cited as the vehicle by which the blue economy finds its *raison d'être* (Patil *et al.*, 2016).

The 2050 Africa's Integrated Maritime Strategy (2050 AIM Strategy) was launched by the AU in 2012, covering many elements of blue economy development (AU, 2012). Since 2014 it has been prominent on the agenda of the Indian Ocean Commission[4] and the Indian Ocean Rim Association (IORA).[5] IORA's Blue Economy Declaration was adopted at the First IORA Ministerial Conference on the blue economy, held in Mauritius in 2015, and promotes the harnessing of ocean and maritime resources to drive economic growth, job creation and innovation, while safeguarding sustainability and environmental protection.[6]

Since then, there have been an increasing number of national and sub-regional initiatives across sub-Saharan Africa that seek to develop blue economy capacity, approaches, strategies and policies, often with support from multilateral development partners such as the World Bank, the UN Development Programme (UNDP), the UN Economic Commission for Africa (UNECA) and the Commonwealth Secretariat. In November 2018 Kenya hosted the world's largest conference to date on the 'Sustainable Blue Economy', with 18,000 registered participants.[7] In the same year Seychelles launched its comprehensive 'Blue Economy Strategy and Roadmap'[8] and President Faure was nominated as the AU blue economy champion. Most recently, the AU launched the Africa Blue Economy Strategy (ABES) in February 2020 during the 33rd AU summit in Addis Ababa, Ethiopia.

The ABES builds on an array of African and international commitments and agreements, providing the basis for a blue economy governance framework. Its objective is to guide the development of an inclusive and sustainable blue economy that becomes a significant contributor to continental transformation and growth. It seeks to achieve this through advancing knowledge on marine and aquatic biotechnology, environmental sustainability, the growth of an Africa-wide shipping industry, the development of sea, river and lake transport, the management of fishing activities on these aquatic spaces, and the exploitation of deep-sea minerals and other resources (AU-IBAR, 2019). The strategy has five themes:

1 Fisheries, aquaculture, conservation and sustainable aquatic ecosystems;
2 Shipping/transportation, trade, ports, maritime security, safety and enforcement;
3 Coastal and maritime tourism, climate change, resilience, environment, infrastructure;
4 Sustainable energy and mineral resources and innovative industries;
5 Policies, institutional and governance, employment, job creation and poverty eradication, innovative financing.

The strategy also highlights 'blue governance and institutional change' as a key strategic challenge, notably for environmental policy, policy coherence and coordination (including marine spatial planning), innovation, accelerated economic transformation and African-led financing mechanisms that support blue growth. Further aspects of governance that are highlighted include developing inclusive approaches to blue economy development (specifically regarding gender, youth and vulnerable people), conflict resolution, awareness raising, transparency and accountability (ibid.).

Missing from ABES but included in the 2050 AIM Strategy is the need to address maritime boundary delimitation. In the 2050 AIM Strategy, states are called on to peacefully resolve existing maritime boundary issues and are encouraged to claim their respective maritime limits and to accept and fulfil their responsibilities emanating from their maritime zones. Conflict avoidance through finalizing maritime boundaries by third party settlement or peaceful cooperation is important for realizing blue economy benefits, particularly where certainty of jurisdiction over resources is concerned.

International and regional frameworks for governing the blue economy

Understanding the range of international and regional instruments on ocean governance, their scope and implementation requirements helps to contextualize contemporary expectations regarding good governance of sustainable blue growth. Covering aspects such as pollution prevention, conservation and sustainable management of resources, many ocean-related agreements over the past three decades have addressed sustainable development and environmental challenges. These remain predominant themes in both international and African agreements that are particularly pertinent to current priorities in blue economy development, including the targets underpinning SDG 14, with its focus on achieving sustainable use of the ocean for economic and other purposes. Explicitly within SDG 14, this is to enhance economic benefits to Small Island Developing States (SIDS) and Least Developed Countries (LDCs) (SDG 14.7), while also ensuring adequate protections are observed: four SDG 14 targets address fisheries and aquaculture (14.4, 14.6, 14.7, 14.b), three ecosystem health, protection and conservation (14.2, 14.5, 14.a) and others pollution (14.1) and the impact of climate change (14.3).

Implementing international law, specifically the 1982 UN Convention on the Law of the Sea, in force in 1994 (henceforth simply UNCLOS) (UNCLOS, 1982), gets its own target (14.c). Forged in the depths of the Cold War, UNCLOS was agreed 10 years ahead of the first UNCSD, reflecting many of the priorities of its day relating to the peaceful use and exploitation of ocean resources, as well as to protection from pollution, and with fair access for all. However, unlike Agenda 21 and the three 'Rio Conventions' that flowed from it,[9] the value of biological diversity has yet to be fully appreciated, and is absent from its text (Tolba, 2008). Over time, as the frontiers of the ocean economy expand, tensions have arisen between these different temporally rooted paradigms, which are also reflected in the founding texts of agreements regarding regional fisheries and seas.

UN Convention on the Law of the Sea

UNCLOS is the primary legal agreement governing the conduct between states in their use of the world's oceans. It is one of the most complex international legal agreements and one of the most fully subscribed (168 states are parties to it). UNCLOS defines the rights and obligations of states in marine waters and the seabed[10] over which their coastal jurisdiction extends; defines the seabed beyond national jurisdiction and its resources as the 'common heritage of [hu]mankind' (UNCLOS, art. 136), and obligates states to protect and preserve the ocean with regard to the rights and jurisdiction of other states. As such, it has often been referred to as the 'Constitution for the Oceans' (Koh, 1982; Zacharias and Ardron, 2020).

As well as defining territorial rights and obligations in the ocean, including the limits of territorial seas, contiguous zones, exclusive economic zones (EEZs) and the continental shelf, UNCLOS covers a wide range of environmentally pertinent themes, *inter alia*:

- delimitation of maritime boundaries;
- fishing, including conservation and management of living marine resources, obligations regarding illegal, unregulated and unreported (IUU) fishing;
- seabed mineral mining, benefit sharing, and protection of the seabed environment;
- environmental protection and prevention of pollution more generally;
- maritime transport, including flag state responsibilities;
- marine scientific research;
- the rights of landlocked, geographically disadvantaged and LDCs (including participation in marine sciences, sharing of fish catches, rights and assistance for access, participation and technology transfer); and
- dispute settlement provisions.

UNCLOS requires global cooperation to manage and protect the marine environment. In particular, states must cooperate to develop science-based rules and standards to maintain environmental quality; protect and preserve rare or fragile ecosystems as well as the habitats of depleted, threatened or endangered species and other forms of marine life; notify other states that are or may be affected by pollution; jointly develop plans to respond to environmental emergencies; undertake joint scientific research; and cooperate on monitoring and environmental assessments. More developed states are obligated to assist less developed states to improve their scientific and research capacity related to the protection of the marine environment (including ensuring that less developed states receive priority funding for these programs). By way of implementation, it is expected that states will adopt and enforce the requirements and principles of UNCLOS in domestic legislation and will cooperate across jurisdictions to prevent and minimize pollution.

Sustainable ocean ecosystems and marine protection

A key tenet of a sustainable blue economy is that durable economic prosperity is dependent on ecological health. The two main targets of SDG 14 seeking to achieve such ecological outcomes are:

- SDG target 14.2: by 2020 to sustainably manage and protect marine and coastal ecosystems to avoid significant adverse impacts, including by strengthening their resilience, and take action for their restoration in order to achieve healthy and productive oceans.
- SDG target 14.5: by 2020 to conserve at least 10% of coastal and marine areas, consistent with national and international law and based on the best available scientific information.

The text of these targets draws on key international agreements, as follows:

The 1992 Convention on Biological Diversity

The key international agreement in relation to holistic protection of biodiversity and ecosystems is the 1992 UN Convention on Biological Diversity (CBD). The three objectives of the Convention are conservation, sustainable use, and the fair and equitable sharing of benefits arising from use of genetic resources. For the marine environment the 1995 Jakarta Mandate on Marine and Coastal Biological Diversity affirms that 'there is a critical need … to address the conservation and sustainable use of marine and coastal biological diversity' (CBD, 2000). Over the past decade, progress towards CBD implementation has been tracked against the Aichi biodiversity targets that were adopted in 2010.

Two key concepts in environmental management highlighted by the CBD that are pertinent to governance of the blue economy are the ecosystem-based approach (EBA) and the precautionary principle or precautionary approach (PP/PA). The EBA concept is the primary framework for action under the CBD. Importantly, it recognizes the holistic nature of ecological systems and that human beings, with their cultural diversity and varied socio-economic endeavours, are an integral component of those systems, thus management approaches also need to be holistic (CBD, 2004).

The PP/PA rose to international prominence through its articulation in Principle 15 of the 1992 Rio Declaration on Environment and Development, which states:

> In order to protect the environment the Precautionary Approach shall be widely applied by States according to their capabilities. Where there are threats of serious or irreversible damage, lack of full scientific certainty shall not be used as a reason for postponing cost-effective measures to prevent environmental degradation.
>
> (UN, 1992)

Essentially, it takes account in decision-making of the vulnerability of the environment, the limitations of science, the availability of alternatives, and the need for long-term, holistic environmental considerations, thereby operating as a safeguard against asymmetric information and imperfect monitoring (Burns, 2007). In a blue economy context, the PP/PA is particularly relevant given widespread scientific uncertainties. It also partially shifts the burden of proof to those promoting an economic activity from those raising ecological objections, although these objections still need to be shown to be plausible and capable of causing irreversible or serious harm. Regulatory approaches such as environmental impact assessments and requirements for pre-development research programmes to address knowledge gaps can be seen as examples of operationalizing the PP/PA, but only if they trigger the use of (cost-effective) measures to prevent environmental degradation. The PP/PA has subsequently been applied widely, for example, as part of the rationale by a national regulator not to consent to a marine mining application (Anton and Kim, 2015), even though this possibility of refusing or halting activities is not explicitly captured in the negotiated text of Rio Principle 15. Therefore, the accepted legal applications of the PP/PA continue to evolve beyond its original formulation.

The SDG 14.5 commitment, to conserve at least 10% of coastal and marine areas by 2020, was derived from target 11 of the CBD's Aichi targets, albeit without the nuances of that original target. From a blue economy perspective, marine protected areas (MPAs) have a range of socio-economic and ecological co-benefits, depending upon the protections in place, such as increased productivity of local fisheries and opportunities for eco-tourism. Aichi target 11 provides a requirement for MPAs to be 'effectively and equitably managed, ecologically representative and well-connected'.[11] The *Atlas of Marine Protection* calculates that 5.3% of the world's oceans are protected in actively managed MPAs, with approximately one-half of that (2.5%) in highly protected no-take marine reserves.[12] The International Union for Conservation of Nature (IUCN) and the World Commission on Protected Areas (WCPA) initiated the IUCN Green List of Protected and Conserved Areas Standard (IUCN and WCPA, 2017) to increase the number of protected and conserved areas that are managed effectively and equitably to deliver conservation outcomes.

The 2003 Revised African Convention on the Conservation of Nature and Natural Resources

For Africa, the key agreement related to the CBD is the 2003 Revised African Convention on the Conservation of Nature and Natural Resources (the 2003 Convention). The original convention, adopted in 1968 in Algiers, Algeria, was considered the most forward-looking regional agreement of its time and it significantly influenced the development of environmental law in Africa (IUCN, 2004). Revised under the auspices of the AU and adopted in 2003 in Maputo, Mozambique, the 2003 Convention was brought up to date with the extensive developments in international environmental law from the intervening 25 years, especially the commitments of the CBD including the PP/PA. It has been recognized by the New Partnership for Africa's

Development (NEPAD) as an essential vehicle for its Environment Action Plan and includes provisions on water pollution and management, conservation and protection of aquatic (including marine) environments, habitats, species and genetic diversity. It also allows for designating marine parks and area-based management tools. However, as is reflected in the 2050 AIM Strategy and ABES, the central African themes of economic development and prosperity are also integral to the 2003 Convention and its focus is more oriented towards sustainable development rather than simply towards conservation. As such, it integrates environmental and socio-economic considerations, for example through provisions for economic incentives and disincentives in order to prevent and abate environmental harm and to restore or enhance environmental quality.[13] In this regard, it is important to note that the Convention limits itself to a definition of natural resources that only includes renewable natural resources. Empowering people through research, education, capacity building and training and strengthening national institutions are also central to the 2003 Convention. Other blue economy-related elements include encouragement to develop and transfer sound environmental technologies.

UNEP Regional Seas Programme, conventions and action plans

Implementing global environmental agreements requires regional context and cooperation, which is what the UN Environment Programme (UNEP) Regional Seas Programme seeks to achieve through its region-based conventions and associated programmes of activity that address environmental and socio-economic aspects of marine and coastal management. Regional seas conventions for sub-Saharan Africa include:

- the 1981 'Abidjan Convention' for Co-Operation in the Protection and Development of the Marine Environment of the West and Central African Region (WACAF);
- the 1982 Regional Convention for the Conservation of the Red Sea and Gulf of Aden Environment ('Jeddah Convention'); and
- the 2010 Amended Nairobi Convention for the Protection, Management and Development of the Marine and Coastal Environment of the Western Indian Ocean ('Nairobi Convention', originally agreed in 1985).

All have associated Action Plans and Protocols to cover aspects such as controlling pollution from a range of sources, emergency responses to pollution, marine protection and biodiversity conservation, sustainable development, environmental assessment and scientific cooperation.

Blue economy development has been identified as a priority by UNEP (UNEP, 2020a) and all the sub-Saharan African regional conventions rely on partnerships with a range of multilateral agencies, governments and development banks. Furthermore, the Nairobi Convention highlights its partnerships with civil society and the private sector in working towards prosperity of the Western Indian Ocean (UNEP, 2020b), creating a range of opportunities for blue economy stakeholders to engage with regional governance bodies.

Maritime security

A key challenge for governing marine waters is the ability to enforce restrictions, such as closed access areas to fishing, over vast swathes of ocean that are largely out of sight and out of mind. Poor management of the ocean and other water bodies is estimated to cost the African region billions of dollars (AU, 2012). Illegal fishing is a particular threat in this regard and is discussed in Chapter 3 in this volume and in more detail in the following section. From an international and regional governance perspective, it is therefore essential for states to be able to take action to ensure the security of their maritime domains. Maritime security issues such as piracy,[14] drug, weapon and human trafficking and terrorism also pose significant threats to countries' abilities to manage their marine estates and develop a sustainable blue economy. Furthermore, these illegal activities are frequently interlinked with illegal fishing and other illegal activities being perpetrated by the same operators and backed by multinational criminal organizations (AU, 2012).

In an effort to tackle this growing threat to Africa's maritime security, during an extraordinary session of the Assembly of the African Union in Lomé, Togo, in 2016, African leaders adopted the African Charter on Maritime Security, Safety and Development, otherwise known as the Lomé Charter (AU, 2016). This charter is consistently aligned with UNCLOS and other international agreements, such as the International Maritime Organization (IMO) Safety of Life at Sea (SOLAS) Convention. Its focus is to protect the region's oceans, seas and waterways from criminal activities, while also advancing safe and equitable exploration that is beneficial to the region and ensuring sustainability of the ocean and its resources. Notably, it gives significant emphasis to a cooperation framework (SOLAS, Chapter V) and provides a comprehensive and inclusive approach to blue economy development (SOLAS, Chapter IV).

In the western Indian Ocean region, further efforts have been made to strengthen cooperation between countries in investigating and arresting those reasonably suspected of piracy, and conducting shared operations. In 2009 the IMO sponsored a meeting in Djibouti, resulting in the Djibouti Code of Conduct. Further amendments were made in Jeddah, Saudi Arabia, in 2017, reiterating the states parties' intention to develop the maritime sector and a sustainable blue economy that generates revenue, employment and stability. All African states of the western Indian Ocean have signed the agreement and have pledged to review their national maritime laws to ensure that they are adequate for criminalizing piracy. One example of regional maritime cooperation that has been helpful in combating these acts is the Regional

Fusion and Law Enforcement Centre for Safety and Security at Sea (REFLECS3), particularly its joint intelligence task force, through which Seychelles, Kenya and Tanzania have worked together to prosecute pirates, thereby restoring confidence in their tourism destinations.

Western Indian Ocean states represented in the IORA have also signed commitments such as the 2017 Jakarta Concord strengthening regional cooperation to address transboundary challenges, including piracy, armed robberies at sea, terrorism, human trafficking, people smuggling, irregular movement of persons, illicit drug trafficking, illicit trafficking in wildlife, crimes in the fisheries sector, and environmental crimes. The Commission of the Economic Community of West African States (ECOWAS), the Economic Community of Central African States and the Gulf of Guinea Commission agreed a Political Declaration, a Memorandum of Understanding and a Code of Conduct regarding piracy and illicit activities in the Gulf of Guinea and adopted the ECOWAS Integrated Maritime Strategy.

Sustainable fishing and fisheries management

Fisheries and aquaculture are a key pillar of any blue economy framework and SDG 14.4 states that the goal is to

> [b]y 2020, effectively regulate harvesting and end overfishing, illegal, unreported and unregulated fishing and destructive fishing practices and implement science-based management plans, in order to restore fish stocks in the shortest time feasible, at least to levels that can produce maximum sustainable yield as determined by their biological characteristics.

Yet progress has fallen well short of this target, with the sustainability of global fishery resources continuing to decline, although at a reduced rate, with the fraction of fish stocks within biologically unsustainable levels sitting at 34% in 2017, up from 10% in 1974 (FAO, 2020).

Fisheries and aquaculture contribute around US $24 billion per year to Africa's economy, representing 1.3% of Africa's total gross domestic product (GDP) in 2011 (de Graaf and Garibaldi, 2014) and engage nearly 60 million people (FAO, 2020). While Africa is home to 20% of the global fishing vessel fleet, around two-thirds of these vessels are unmotorized and 90% are less than 12 m in length, reflecting the fact that the vast majority of fishers operate in small-scale fisheries. Mauritania is the only sub-Saharan African country in the top 25 global fish producers and contributes 1% of the global total fish production, due largely to extensive agreements with distant water fishing nations (ibid.). Yet 25% of all marine catches around Africa are by non-African countries and could add an additional $3.3 billion to the African economy if they were caught by African vessels, which is eight times higher than the $0.4 billion that African countries earn from fisheries agreements (de Graaf and Garibaldi, 2014). It is estimated that a further $2–$10 billion is lost annually due to mismanagement and unsustainable practices (AUC-NEPAD, 2014; World Bank, 2017). Thus, effective governance through improved management mechanisms and policy implementation is a priority for the success of blue economy development and countries that have made efforts in this regard have seen tangible benefits. For example, Namibia's commitment to long-term fisheries governance, enforcement and policy implementation[15] has resulted in a 15% decline in overexploited and collapsed fish stocks over a six-year period. Meanwhile, Tanzania's deployment of beach management units has increased community involvement in the management and surveillance of local fisheries (UNEP, 2019).

Addressing this challenge requires having an understanding of the range of international and regional governance arrangements that apply to marine capture fisheries. Setting the framework for these is UNCLOS, which covers various diverse aspects related to fishing, including the conservation and management of living marine resources, obligations of states regarding illegal, unregulated and unreported (IUU) fishing, and sharing of fish catches and access arrangements to fisheries with landlocked, geographically disadvantaged and LDCs. Subsequent agreements include both 'hard law' and 'soft law' agreements. 'Hard law' agreements include the 1993 Food and Agricultural Organization of the UN (FAO) Agreement to Promote Compliance with International Conservation and Management Measures by Fishing Vessels on the High Seas (FAO Compliance Agreement), which entered into force in 2003; the 1995 Agreement for the Implementation of the Provisions of UNCLOS Relating to the Management of Straddling Fish Stocks and Highly Migratory Fish Stocks (UN Fish Stocks Agreement – UNFSA) that entered into force in 2001, and more recently the 2009 Port State Measures Agreement (PSMA) which entered into force in 2016. 'Soft law' agreements include the 1995 FAO Code of Conduct for Responsible Fisheries and its associated FAO international plans of actions (IPOAs) and the various resolutions arising from the 1998 Jakarta Mandate under the CBD. The UN specialist agency relating to fisheries is FAO, through its Fisheries and Aquaculture Department and this has responsibility for the administration and implementation of these agreements and associated IPOAs (Zacharias and Ardron, 2020).

Ecosystem Approach to Fisheries

A key concept underpinning modern fisheries management is the Ecosystem Approach to Fisheries (EAF), defined by FAO in the 2001 Reykjavik Declaration on Responsible Fisheries:

> An ecosystem approach to fisheries strives to balance diverse societal objectives, by taking into account the knowledge and uncertainties about biotic, abiotic, and human components of ecosystems and their

> interactions and applying an integrated approach to fisheries within ecologically meaningful boundaries.
>
> (FAO, 2003)

The EAF provides some alignment between concepts in fisheries management and those underpinning biodiversity conservation (the EBA of the CBD) and promotes integrated planning. Fundamentally, it recognizes that successful fisheries management relies not just on managing a selection of commercially important populations but on taking account of the environment in which they live and the activities of human beings as a key component of that wider environment.

Regional fisheries management bodies

Around the time that UNCLOS came into force, the need to better regulate transboundary fisheries was also being recognized. Subsequently, the UN Fish Stocks Agreement with its focus on highly migratory and straddling stocks mandated the establishment of Regional Fisheries Management Organizations (RFMOs). Through these, states are to cooperate to manage fisheries and to set legally binding quotas for Areas Beyond National Jurisdiction, as well as sometimes within national jurisdictions, depending on the nature of the agreement. Several subregional fisheries committees and fisheries advisory bodies also exist (Zacharias and Ardron, 2020).

RFMOs, like most regional agreements, lack supranational powers, meaning that states have to agree to be bound by RFMO decisions. This is one reason why some question whether RFMOs are meeting their conservation mandates (Cullis-Suzuki and Pauly, 2010). Certain RFMOs, particularly those established to manage lucrative tuna fisheries, have been criticized for ignoring scientific advice and setting harvest levels that are incompatible with an EBA (Gilman *et al.*, 2014; Juan-Jordá *et al.*, 2015).

Large Marine Ecosystems

An alternative and complementary approach to regional fisheries management emerged during the 1980s that considered the need to manage living marine resources at the natural spatial scale of their wider ecosystem, acknowledging that the ocean and its biotic inhabitants move over ranges that do not respect political borders (Sherman and Alexander, 1986). 'Large Marine Ecosystems' (LMEs) combined earlier thinking on EBAs with the recognition that regional water bodies and ocean current systems provided a more scientifically appropriate scale to manage living marine resources, and have led to developing holistic ecosystem-based measures of overharvesting fish not captured by conventional fisheries management (Link *et al.*, 2020). Globally, 66 LMEs have been recognized, each defined by unique undersea topography, current dynamics, marine productivity and trophic interactions and named after the defining ocean feature of their domain, usually a large ocean current or regional sea (Sherman and Alexander, 1986; Hudson, 2017).

The LME approach has been particularly well received by African states. Within the waters around sub-Saharan Africa, six LMEs have been recognized: the Canary Current; the Guinea Current; the Benguela Current; the Agulhas Current; the Somali Coastal Current and the Arabian Sea. All except the Arabian Sea LME have established Strategic Action Programmes that provide support for policy harmonization and institutional reforms, together with Transboundary Diagnostic Analysis that addresses environmental and management issues within the context of the multi-jurisdictional status of the waters. Some LMEs have sought to establish National Intersectoral Coordination mechanisms to coordinate between the wide range of stakeholders engaged with governance of the LME at the national and regional level (McConney *et al.*, 2016). The Guinea Current and Benguela Current LME programmes have both resulted in establishing interim regional commissions as recognized international bodies. The commissions serve to promote a coordinated regional approach to the long-term conservation and environmentally sustainable development of the LME. In the case of the Benguela Current Commission, its status was made permanent by the signing of the Benguela Current Convention in 2013.

The Pan-African Fisheries and Aquaculture Policy Framework and Reform Strategy

As can be seen from the above range of fisheries policy instruments, management organizations and advisory bodies governing fisheries around the African continent, it is a complex and somewhat overlapping space. To overcome the human and institutional capacity constraints in the region requires major governance reforms in national fisheries administrations and regional arrangements for fisheries management (AUC-NEPAD, 2014).

To this end, African fisheries governance instruments have developed substantially over the past two decades, from the 2005 Fish for All Summit organized by the AU and NEPAD in Abuja, Nigeria – where African Heads of State and Government endorsed NEPAD's Action Plan for the Development of African Fisheries and Aquaculture – through to the first Conference of African Ministers of Fisheries and Aquaculture[16] and its subsequent endorsement[17] as the policy organ responsible for fisheries and aquaculture within the Conference of African Ministers of Agriculture, to adopting the AU/NEPAD Policy Framework and Reform Strategy for Fisheries and Aquaculture in Africa in 2014 (AUC-NEPAD, 2014).

The Pan-African Fisheries and Aquaculture Policy Framework and Reform Strategy (PFRS) takes into consideration specific regional priorities of common interest across Africa, the supporting and delivery mechanisms needed to assist and facilitate implementation of the agreed strategies and measures to track success. It has provided a template for harmonizing policy, encompassing marine fisheries, lake fisheries, aquaculture and river-based/inland fisheries in the form of, *inter alia*, continental and national roadmaps, manuals and model agreements.[18] Policy statements in these areas often reflect the aims of the PFRS as a framework directive.[19] As the PFRS was established prior to the ABES, policy direction under the PFRS incorporates elements from FAO guidelines, e.g. IUU fishing and knowledge gleaned from regional stakeholder consultations, training and cooperation.[20] This provides useful guidance for implementing similar aspects under the ABES.

Illegal, unreported and unregulated (IUU) fishing

IUU fishing has significant adverse impacts on economic flows and management efforts in fisheries. Globally, the cost of IUU fishing has been estimated between US $10–$24 billion per year (Agnew *et al.*, 2009). In West Africa, the problem is particularly acute with IUU fishing potentially accounting for an estimated 40% of all fish caught (ibid.), costing the region $2.3 billion per year and over 300,000 jobs (Belhabib, 2017).

Two global agreements seeking to tackle the market and vessel aspects of IUU fishing have either come into force recently or are close to doing so and, when implemented together, will provide a stronger framework for tackling IUU fishing. These are the FAO's Port State Measures Agreement (PSMA, 2009, in force in 2016), and the IMO's Cape Town Agreement (CTA, 2012, not yet in force but with a target date of 2022).

The PSMA is the first binding international agreement that specifically targets IUU fishing.[21] The agreement provides a framework to strengthen port controls on vessels landing catches, and to prevent illegally caught fish from entering the global market. The PSMA specifies national-level integration and coordination, international cooperation and exchange of information, designated ports where inspections will be carried out and, most significantly, the ability to deny a vessel entry to a port and its provisioning services if the vessel does not have authorization to fish or there is evidence that it has been involved in IUU fishing.

The IMO's CTA consists of minimum safety measures for fishing vessels that mirror the SOLAS Convention. It also calls for harmonized fisheries, labour and safety inspections with other agreements.[22] Once in force, the CTA will set minimum requirements on the design, construction, equipment and inspection of fishing vessels that are 24 m in length or longer that operate on the high seas (Pew Charitable Trusts, 2020), with significant implications for standards within the commercial fishing industry. The inspection element of this agreement will particularly help to strengthen the legal apparatus to tackle IUU fishing.

Marine plastic pollution

Marine plastic pollution has been the recent focus of much of the international ocean governance agenda and is addressed generally in SDG target

14.1, which commits that it will 'By 2025, prevent and significantly reduce marine pollution of all kinds, in particular from land-based activities, including marine debris and nutrient pollution'. Pollution provisions in UNCLOS were fuelled by two large marine oil spills from ships: the *Torrey Canyon* in 1967 and the *Amoco Cadiz* in 1978, reinforced by subsidiary agreements of the IMO[23] and UNEP (the CBD and Regional Seas Programme) and through African agreements such as the 2009 Revised African Convention on the Conservation of Nature and Natural Resources.

What was not envisaged when these earlier agreements were drafted was the forthcoming ubiquitous prevalence of plastic pollution across the planet, even in the remotest Arctic habitats and at the bottom of the deepest ocean trenches. Plastic pollution is now recognized as one of the greatest pollution threats to marine ecosystems and human well-being (Beaumont *et al.*, 2019) estimated the global societal cost of marine plastic pollution to be in the region of US $500–$2,500 billion per year. While most plastic pollution in the ocean originates from terrestrial sources, a significant proportion also arises from discarded fishing gear that can continue to catch marine life after it has been abandoned, causing lasting ecological and socio-economic damage (Gilman, 2015; Scheld *et al.*, 2016).

The 1989 Basel Convention on Control of Transboundary Movements of Hazardous Wastes and their Disposal includes recyclable materials and was amended in 2019 to include plastic waste, in order to make its global trade more transparent, better regulated and more safely managed. Persistent organic pollutants contained in plastic waste, such as flame retardants and plasticizers, are covered by the 2001 Stockholm Convention.

Nonetheless, despite the existence of some international provisions, the increasing scale of the problem and the failure to mount any legal challenge to polluting states under these international regimes shows the insufficiency of the current framework to deal with these challenges. Filling this policy gap, a number of voluntary global intergovernmental agreements on plastic pollution have launched in the past few years, such as the 2018 G7 Ocean Plastics Charter[24] and the 2018 Commonwealth Clean Ocean Alliance of the Commonwealth Blue Charter,[25] through which 34 countries to date, including 14 from Africa, have committed to take domestic action to tackle single-use plastics and plastic micro-beads in cosmetic products. How these will influence the growing calls for a binding international treaty on plastic pollution is yet to be seen.

Climate change and the blue economy

Anthropogenic climate change is probably the greatest governance challenge the world has ever faced. The role of the ocean in mitigating and adapting to climate change has been well documented (e.g. IPCC, 2019), together with its implications for the ocean economy and livelihoods (e.g. Gaines *et al.*, 2019; Barange *et al.*, 2018). Certainly, the impact of climate change can already be seen across Africa, with the continent experiencing more than 2,000 natural disasters since 1970, almost half of which have occurred in the past decade (World Bank, 2019).

Despite its importance, recognition of the role of the ocean in climate mitigation and adaptation has been weak. The United Nations Framework Convention on Climate Change (UNFCCC) text commits states parties to 'Promote sustainable management, and promote and cooperate in the conservation and enhancement, as appropriate, of sinks and reservoirs of all greenhouse gases … including biomass, forests and oceans as well as other terrestrial, coastal and marine ecosystems' (art. 4 1(d)).[26] The preamble to the Paris Agreement notes 'the importance of ensuring the integrity of all ecosystems, including oceans, and the protection of biodiversity'.[27] The Intergovernmental Panel on Climate Change (IPCC) has increasingly highlighted the role of the ocean as a major carbon sink, as well as the impact it is facing from climate change, including from ocean acidification (IPCC, 2019). Nonetheless, the focus to date has largely ignored ocean-related mitigation strategies, compared to terrestrial forests and the UN Programme on Reducing Emissions from Deforestation and Forest Degradation (REDD) and REDD+ initiatives, for example.

The Paris Agreement's main implementation vehicle is Nationally Determined Contributions (NDCs), whereby countries commit to ambitious but voluntary levels of greenhouse gas reduction. Adaptation measures can also be included in NDCs. These are particularly important for many developing countries with low emissions and high exposure to the impact of climate change. Countries can also develop National Adaptation Plans (NAPs) to help to build climate resilience. Countries can include ocean-related measures both for adaptation (e.g. strengthening shoreline defences through protection of coastal habitats) and mitigation (e.g. developing blue carbon stocks by restoring mangrove forests) in the NDCs and NAPs.

In the context of the blue economy, the role of 'natural capital assets' in adapting to climate change has been highlighted as a key area for investment that can be included in NDCs and NAPs. Coastal ecosystems, such as coral reefs, mangroves and salt marshes, can provide natural sea defences by mitigating the impact of storm surges, hurricanes and tsunamis, while also supporting coastal fisheries. Protecting and restoring such ecosystems will build climate resilience and provide for coastal livelihoods. Furthermore, blue carbon coastal ecosystems, such as mangroves and seagrasses, sequester carbon dioxide (CO_2) emissions with per metre efficiencies greater than those of most terrestrial ecosystems. Thus, they can be included as part of NDC mitigation plans and could earn credits through (expanded) carbon trading systems, if validated through a recognized verification scheme. Such schemes are still in their infancy and will require concerted actions by governments to mainstream their usage.

Developing ocean-based renewable energy would reduce reliance on fossil fuel consumption, particularly so for SIDS, and may be included in NDCs and NAPs. Offshore wind already makes a strong contribution to energy generation in developed countries, providing more than one-half of the United Kingdom's energy needs, for example (UK BEIS, 2020). Although still in development, wave, tidal, floating solar and ocean thermal energy conversion systems also show promise (Haugan *et al.*, 2020).

The largest source of greenhouse gas emissions in the blue economy comes from shipping, with emissions equating to around 2.5% of the global total (IMO, 2015), approximately equivalent to the total energy sector emissions from sub-Saharan Africa, and more than that produced by the global aviation industry.[28] Owing to growth in international trade, shipping emissions are projected to increase by between 50% and 250% by 2050 (ibid.). Emissions from shipping are not restricted to CO_2. Other air pollutants include sulphur oxides (SOx), nitrogen oxides (NOx), and fine particles (black carbon), all of which affect the global climate (and human health in coastal regions).

In 2018 the IMO set out its strategy for reducing total greenhouse gas emissions from ships, with a view to reach the peak in emissions as soon as possible, by 2030 to reduce the carbon intensity of shipping by 40% and by 2050 to have decreased annual emissions by 50% compared to 2008 levels, while also pursuing efforts to phase them out entirely (IMO Resolution MEPC.304(72)). Mandatory implementation of these targets is now being proposed through amendments to the MARPOL Convention. These targets have been criticized for falling short of the levels of reduction needed to achieve the Paris Agreement.[29] Other amendments to MARPOL include new regulations that entered into force in 2020 aimed at significantly reducing SOx emissions from shipping (IMO, Resolution MEPC.280(70)).

Reconciling the stated ambitions of African blue economy strategies to develop maritime-based trade and shipping[30] with national commitments under the Paris Agreement will require an equally strong focus on greener shipping, for example through cleaner fuels and renewable energy propulsion systems, as well as from more energy-efficient port infrastructure – a focus not yet included into these strategies.

Deep seabed mining

Deep seabed mining (DSM) refers to the extraction of mineral deposits from the deep seafloor, including many metals that are considered essential for low-carbon technologies and the sustainable energy transition (e.g. batteries for renewable energy storage). DSM has not yet occurred anywhere. In areas beyond national jurisdiction (ABNJ, also known as the Area), it was a major focus of the final round of UNCLOS negotiations from 1973 to 1982. International regulation of DSM in the Area is by the International Seabed Authority (ISA), established under UNCLOS and the resources of the seabed in the Area are regarded as the 'common heritage of [hu]mankind' (UNCLOS, art. 136). The ISA has responsibility for awarding contracts to explore for, or exploit, seabed minerals in the Area and has granted 30 exploration contracts. However, the development of exploitation regulations is still under negotiation and their implementation of sustainability principles remains to be seen. A contract can be granted directly to a state party to UNCLOS, or to an entity (private or state enterprise) sponsored by a state party.

There are currently no sponsoring states from Africa, which has been described as 'rather surprising' given that the perceived potential African participation could contribute to the sustainable development of the continent (Egede, 2017). A number of key African blue economy strategic documents have identified DSM as a key industry to develop for African prosperity. They include, most importantly, the ABES and AU Agenda 2063, but also the 1988 Kampala Programme of Action on the Development and Utilisation of Mineral Resources in Africa[31] (UNECA, 1988). UNECA (2016) has welcomed the idea of African participation in seabed mining in the Area, observing that it is in the strategic interest of Africa and its blue economy development. The ABES highlights the importance of DSM to Africa's economic prosperity and proposes 2020–23 as the timeframe for African states to develop appropriate national regulatory frameworks (AU-IBAR, 2019). However, the ABES focuses on DSM within Africa's EEZ and is practically silent on Africa's participation in DSM in the Area.

It is important to note in this discussion that the potential economic returns that DSM is speculated to generate have yet to be validated given that exploitation regulations and the payment regime have not yet been agreed at the ISA and no historical economic data exist for the industry. Furthermore, attempts to develop DSM in non-African national jurisdictions, most recently in Papua New Guinea, have not been financially successful.[32] This, together with the continent's stakes in terrestrial mining, may go some way to explaining African nations' reticence to become sponsoring states (for possible explanations of this reticence, and the pros and cons of African states becoming sponsoring states, see Nwapi and Wilde, 2020), highlighting a possible disconnect between continent-wide strategies and national ambitions. In addition, and similarly to elsewhere in the world, there are concerns regarding the current capacity of African states to monitor seabed miners to ensure that they comply with their environmental and fiscal obligations (Van Nijen *et al.*, 2019; Navarre and Lammens, 2017).

Conclusion

This chapter has summarized some of the key African and global multilateral agreements that underpin (or should underpin) governance of the blue economy in Africa. It highlights the marine environmental components of these agreements in particular because economic activities that occur in or on the ocean itself are most likely to have a direct bearing on the marine environment and require their own instruments to minimize and balance economic development with environmental sustainability. Nonetheless, social and economic governance arrangements are of equal importance and should also be incorporated into blue economy frameworks.

Tackling challenges such as infrastructure investments, international trade, illicit financial flows and organized crime in the blue economy requires full consideration of a suite of other international and regional governance arrangements to ensure the development of the appropriate frameworks. For example, a constraint in dealing with IUU fishing has been the tendency to treat it as a fisheries management problem rather than as a criminal enforcement issue. Attempts to redefine IUU fishing as fisheries crime in South Africa have shown the benefits of harmonizing across institutional arrangements to tackle the problem (de Coning and Witbooi, 2015).

A summary of key ocean governance instruments such as this one is insufficient to elucidate major gaps in global and regional ocean governance but gaps do exist and will become increasingly problematic as blue economy development gains pace. These gaps are not only in legislation but also in the implementation and enforcement of existing laws and agreements. 'Environmental rule of law' requires well-designed laws implemented by capable government institutions held accountable by an informed, engaged public. Furthermore, to track regional (and global) progress, a set of indicators consistent across national jurisdictions is required (UNEP, 2019). Without the translation of existing international agreements into national laws, policies and regulations that are implemented, the hortatory goals expressed by this constellation of international ocean governance arrangements will remain out of reach to most African countries. Barriers to implementation are numerous and vary from place to place, but in all cases a failure to have robust environmental institutions can create 'a system of broader institutional weakness which can result in corruption' (Kaufmann, cited in UNEP, 2019, p. 31). Regional cooperation mechanisms and cooperative approaches between countries addressing similar challenges provide a means to overcome these barriers. The number of states that are actively cooperating with one another to tackle shared ocean issues is on the rise; for example, the Commonwealth Blue Charter has more than 40 countries participating in one or more of its 10 Action Groups, and provides a model of how country-led platforms for knowledge exchange, peer learning, capacity building and partnership can help to overcome national silos and challenges (Ardron *et al.*, 2020).

Tensions between paradigms of economic growth and sustainability are not unique to the ocean or to sub-Saharan Africa. The 'Blue Growth' agenda in Europe, for example, has spawned research pointing out the (rather obvious) conclusion that blue economies require healthy ecosystem services to persist (Lillebø *et al.*, 2017). Furthermore, when wrongly implemented, blue growth can lead to 'ocean grabbing' that displaces local interests (Barbesgaard, 2018; Bennet *et al.*, 2015). Indeed, it could be said that the African experience has been similar to the European one in this and other regards (c.f. Okafor-Yarwood *et al.*, 2020). Therefore, Africa is not alone in its struggles to strike the right balance. However, without harmonizing policy objectives across ministries and states, the tensions of conflicting policy agendas will persist. The focus on oil and gas as a blue economy sector in ABES, for example, would appear to be contrary to international commitments of all sub-Saharan African states under the UNFCCC Paris Agreement. Promoting economic growth in a potentially lucrative industry already in place elsewhere is an understandable aspiration, but at this late stage in the global energy economy comes with the very real risk of stranded capital investments and assets.

Caswell *et al.* (2020) have highlighted several cross-cutting lessons for blue growth, drawing on historical perspectives in fisheries governance. Growth in the blue economy needs to have a clear understanding of its ecological and social limits and should be assessed over a range of scales and time frames, not prioritizing short-term gains that can lead to long-term losses in growth. Equitable approaches underpinned by scientific knowledge and embedded within an adaptive and holistic governance framework will have the highest chance of success.

Ultimately, the touchstone of success in implementing the blue economy must not be solely the year's economic returns, but rather the rate of transition from unsustainable to sustainable activities that will yield returns over the much longer term. Ocean economic output is forecast to double by 2030 (OECD, 2016) and too often this assertion alone has been used as sufficient justification for the blue economy concept. Given that the biggest threat to this euphoric vision of endless blue growth is environmental degradation and the subsequent loss of ecosystem services (Lillebø *et al.*, 2018), it follows that such a threat should be more closely examined. Curtailing the unsustainable harvesting of marine resources, recognizing the natural capital value of coastal resources, and better enforcing against polluting activities within the shared ocean, must remain at the heart of all blue economy governance –in sub-Saharan Africa and globally. The agreements outlined in this chapter have the best of intentions but are powerless without national implementation. Only then will the true potential of the blue economy be realized.

Notes

1 The green economy concept was originally promoted through the UN Conference on Environment and Development held in Rio de Janeiro, Brazil, in 1992, but gained momentum again after the global financial crisis of 2008. The blue economy concept gained popularity in the run-up to the UNCSD in 2012, largely championed by Small Island Developing States.

2 www.un.org/sustainabledevelopment/oceans/.

3 Aichi targets of the Convention on Biological Diversity; Sustainable Development Goal 14 early targets (UN, Progress towards the Sustainable Development Goals: Report of the Secretary-General, 2020).

4 The Indian Ocean Commission comprises Indian Ocean island states in Africa, namely Comoros, Madagascar, Mauritius, Seychelles and Réunion.

5 While not specifically an African organization, IORA includes all East African coastal states between Somalia and South Africa, as well as Indian Ocean islands of Africa. The blue economy was championed by South Africa during the 14th Ministerial Meeting of IORA, in Perth, Australia, in 2014.

6 A subsequent IORA Ministerial Conference on the blue economy was held in Jakarta, Indonesia, in 2017 and resulted in the Jakarta Declaration that focused on financing the blue economy.

7 This followed Kenya stepping forward as Commonwealth Champions on the blue economy at the launch of the Commonwealth Blue Charter during the Commonwealth Heads of Government Meeting in April 2018.

8 A foundational document for the AU's Africa Blue Economy Strategy.

9 The three Rio Conventions – the Convention on Biological Diversity, the United Nations Convention to Combat Desertification and the United Nations Framework Convention on Climate Change – have their roots in the 1992 Earth Summit (https://www.cbd.int/rio/).

10 As well as the overlying airspace.

11 www.cbd.int/aichi-targets/target/11.

12 www.mpatlas.org/.

13 Economic incentives are also recognized in the CBD.

14 The Gulf of Guinea region is the dominant region globally for piracy, with 95% of global kidnappings at sea occurring in this region (ICC-IMB, 2020).

15 Including high penalties for illegal vessels caught within its maritime jurisdiction and a monitoring system covering 91.5% of all seagoing vessels aimed at preventing illegal fishing.

16 Held in Banjul, The Gambia, in 2010.

17 By the 18th Session of the AU Assembly of Heads of State, in 2011.

18 Best Practices and Guidelines to support Commercial Aquaculture Enterprise Development in Africa; www.au-ibar.org/component/jdownloads/finish/76-tmt/3271-best-practices-and-guidelines-to-support-commercial-aquaculture-enterprise-development-in-africa.

19 See Pan-African Workshop on Strengthening Organizational Structures of Non-State Actors for Sustainable Small-Scale Fisheries in Africa in Botswana 2019; www.fao.org/3/ca8141en/CA8141EN.pdf..

20 The PFRS reflects policy direction from FAO Guidelines (S.1.2.1). Training, think tanks and stakeholder consultations are referenced throughout S.1.4 of the PFRS.

21 www.fao.org/port-state-measures/en/.

22 Including a third related agreement on the working environment of fishers: the 2007 International Labour Organization's Work in Fishing Convention No. 188 (C188, in force in 2017), which may help to address some of the human exploitation aspects of IUU fishing. Further discussion of C188 lies outside the main focus of this chapter; nonetheless, its provisions on decent work conditions for the commercial fishing industry are of strong relevance to blue economy governance.

23 The International Convention for the Prevention of Pollution from Ships (MARPOL 73/78) and its six annexes cover pollution arising from ships and offshore installations. The 1972 London Convention/1996 London Protocol (LC/LP) regulates dumping activities at sea.

24 www.consilium.europa.eu/media/40516/charlevoix_oceans_plastic_charter_en.pdf.

25 https://bluecharter.thecommonwealth.org/action-groups/marine-plastic-pollution/.

26 https://unfccc.int/resource/docs/convkp/conveng.pdf.

27 https://unfccc.int/files/meetings/paris_nov_2015/application/pdf/paris_agreement_english_.pdf.

28 The global aviation industry is estimated to produce 2% of CO_2 emissions; www.atag.org/facts-figures.html.

29 www.theguardian.com/environment/2020/oct/23/green-groups-condemn-proposals-to-cut-shipping-emissions.
30 2050 AIM Strategy Objective IV and ABES Thematic Area 2.
31 Adopted during the Third Regional Conference on the Development and Utilisation of Mineral Resources in Africa.
32 www.theguardian.com/world/2019/sep/16/collapse-of-png-deep-sea-mining-venture-sparks-calls-for-moratorium.

References

African Union (AU) (2012). *2020 Africa's Integrated Maritime Strategy (2050 AIM Strategy)*. Addis Ababa: African Union.

African Union (AU) (2015). *Agenda 2063: The Africa We Want*. Addis Ababa: African Union.

African Union (AU) (2016). *African Charter on Maritime Security and Safety and Development in Africa (Lomé Charter)*. Addis Ababa: African Union. https://au.int/sites/default/files/treaties/37286-treaty-african_charter_on_maritime_security.pdf.

African Union Commission-New Partnership for Africa's Development (AUC-NEPAD) (2014). *The Policy Framework and Reform Strategy for Fisheries and Aquaculture in Africa*. Addis Ababa: African Union.

African Union Inter-African Bureau for Animal Resources (AU-IBAR) (2019). *Africa Blue Economy Strategy.* Nairobi: African Union.

Agnew, D. J., Pearce, J., Pramod, G., Peatman, T., Watson, R., Beddington, J. R. and Pitcher, T. J. (2009). 'Estimating the Worldwide Extent of Illegal Fishing'. *PLOS ONE*, 4, e4570. doi:doi:10.1371/journal.pone.0004570.

Anton, D. K. and Kim, R. E. (2015). 'The Application of the Precautionary and Adaptive Management Approaches in the Seabed Mining Context: Trans-Tasman Resources Ltd Marine Consent Decision under New Zealand's Exclusive Economic Zone and Continental Shelf (Environmental Effects) Act 2012'. *The International Journal of Marine and Coastal Law*, 30(1), 175–188.

Ardron, J. A., Swaddling, A. A., Prislan, H. and Hardman-Mountford, N. (2020). 'Sailing on an Ocean of Noble Causes: The Commonwealth Blue Charter'. *Ocean Yearbook Online*, 34(1), 1–19. doi:doi:10.1163/9789004426214_002.

Barange, M., Bahri, T., Beveridge, M., Cochrane, K., Funge-Smith, S. and Poulain, F. (eds) (2018). *Impacts of Climate Change on Fisheries and Aquaculture: Synthesis of Current Knowledge, Adaptation and Mitigation Options.* Rome: FAO.

Barbesgaard, M. (2018). 'Blue Growth: Savior or Ocean Grabbing?' *The Journal of Peasant Studies*, 45(1), 130–149.

Beaumont, N. J., Aanesen, M., Austen, M. C., Börger, T., Clark, J. R., Cole, M., Wyles, K. J. (2019). 'Global Ecological, Social and Economic Impacts of Marine Plastic'. *Marine Pollution Bulletin*, 142, 189–195.

Belhabib, D. (2017). 'West Africa: Illegal Fishing, The Black Hole in the Seas'. *Samudra Report*, 77, 20–25. www.icsf.net/images/samudra/pdf/english/issue_77/4319_art_Sam77_e_art06.pdf.

Bennet, N., Govan, H. and Satterfield, T. (2015). 'Ocean Grabbing'. *Marine Policy*, 57, 61–68.

Burns, W. C. (2007). 'Potential Causes of Action for Climate Impacts under the United Nations Fish Stocks Agreement'. *Sustainable Development Law & Policy* (Winter), 34–38, 81–82.

Caswell, B., Klein, E., Alleway, H., Ball, J., Botero, J., Cardinale, M., … Hentati-Sundberg, J. (2020). 'Something Old, Something New: Historical Perspectives Provide Lessons for Blue Growth Agendas'. *Fish and Fisheries*, 21(4), 774–796. doi: doi:10.1111/faf.12460.

Convention on Biological Diversity (CBD) (2000). *The Jakarta Mandate: From Global Consensus to Global Work*. www.cbd.int/doc/publications/jm-brochure-en.pdf (accessed 19 November 2020).

Convention on Biological Diversity (CBD) (2004). *The Ecosystem Approach*. Montreal: Secretariat of the Convention on Biological Diversity. www.cbd.int/doc/publications/ea-text-en.pdf.

Cullis-Suzuki, S. and Pauly, D. (2010). 'Failing the High Seas: A Global Evaluation of Regional Fisheries Management Organizations'. *Marine Policy*, 34, 1036–1042.

de Coning, E. and Witbooi, E. (2015). 'Towards a New "Fisheries Crime" Paradigm: South Africa as an Illustrative Example'. *Marine Policy*, 208–215. doi:doi:10.1016/j.marpol.2015.06.024.

de Graaf, G. and Garibaldi, L. (2014). 'The Value of African Fisheries'. *FAO Fisheries and Aquaculture Circular* (1093), 76.

Department of Business, Energy and Industrial Strategy (BEIS) (2020). *Digest of United Kingdom Energy Statistics*. London: Department of Business, Energy and Industrial Strategy. https://assets.publishing.service.gov.uk/government/uploads/system/uploads/attachment_data/file/924591/DUKES_2020_MASTER.pdf.

Egede, E. (2017). 'A "New Frontier" for Mining? Time for Africa's Engagement with Deep Seabed Mining'. *International Bar Association Public Law Committee Newsletter*. www.ibanet.org/Article/NewDetail.aspx?ArticleUid=AFDB69A8-7C83-4184-91B4-2A814556BEF7.

Food and Agricultural Organization of the United Nations (FAO) (2020). *The State of World Fisheries and Aquaculture 2020: Sustainability in Action*. Rome: FAO. doi:10.4060/ca9229en.

Gaines, S., Cabral, R., Free, C., Golbuu, Y.*et al.* (2019). *The Expected Impacts of Climate Change on the Ocean Economy*. Washington, DC: World Resources Institute.

Gilman, E. (2015). 'Status of International Monitoring and Management of Abandoned, Lost and Discarded Fishing Gear and Ghost Fishing'. *Marine Policy*, 60, 225–239.

Gilman, E., Passfield, K. and Nakamura, K. (2014). 'Performance of Regional Fisheries Management Organizations: Ecosystem-Based Governance of Bycatch and Discards'. *Fish and Fisheries*, 15(2), 327–351.

Haugan, P., Levin, L., Amon, D., Hemer, M., Lily, H. and Nielsen, F. (2020). *What Role for Ocean-Based Renewable Energy and Deep Seabed Minerals in a Sustainable Future?*Washington, DC: World Resources Institute. www.oceanpanel.org/blue-papers/ocean-energy-and-mineral-sources.

Hudson, A. (2017). 'Restoring and Protecting the World's Large Marine Ecosystems: An Engine for Job Creation and Sustainable Economic Development'. *Environmental Development*, 22, 150–155.

Intergovernmental Panel on Climate Change (IPCC) (2019). *IPCC Special Report on the Ocean and Cryosphere in a Changing Climate*. Geneva: IPCC.

International Chamber of Commerce- International Maritime Bureau (ICC-IMB) (2020). *Piracy and Armed Robbery Against Ships*. London: ICC-International Maritime Bureau.

International Maritime Organization (IMO) (2015). *Third IMO Greenhouse Gas Study 2014*. London: International Maritime Organization.

International Union for Conservation of Nature (IUCN) and World Commission on Protected Areas (WCPA) (2017). *IUCN Green List of Protected and Conserved Areas: Standard, Version 1.1*. Gland: IUCN.

International Union for Conservation of Nature (IUCN) (2004). *An Introduction to the African Convention on the Conservation of Nature and Natural Resources*. Gland and Cambridge: IUCN.

Johansen, D. and Vestvik, R. (2020). 'The Cost of Saving Our Ocean: Estimating the funding Gap of Sustainable Development Goal 14'. *Marine Policy*, 112.

Juan-Jordá, M., Arrizabalaga, H., Restrepo, V., Dulvy, N., Cooper, A. and Murua, H. (2015). *Preliminary Review of ICCAT, WCPFC, IOTC and IATTC Progress in Applying Ecosystem Based Fisheries Management*. Indian Ocean Tuna Commission (IOTC) Technical Report.

Kaufmann, D. (2015). 'Evidence-Based Reflections on Natural Resource Governance and Corruption in Africa'. In E. Zedillo, O. Cattaneo and H. Wheeler (eds), *AFRICA at a Fork in the Road* (p. 239). Newhaven, CT: Yale Center for the Study of Globalization.

Koh, T. (1982). *Remarks by Tommy Koh, President of the Third United Nations Conference on the Law of the Sea*. www.un.org/depts/los/convention_agreements/texts/koh_english.pdf (accessed 20 November 2020).

Lillebø, A., Pita, C., Rodrigues, J. G., Ramos, S. and Villasante, S. (2017). 'How Can Marine Ecosystem Services Support the Blue Growth Agenda?' *Marine Policy*, 132–142. doi:doi:10.1016/j.marpol.2017.03.008.

Link, J., Watson, R., Pranovi, F. and Libralato, S. (2020). 'Comparative Production of Fisheries Yields and Ecosystem Overfishing in African Large Marine Ecosystems'. *Environmental Development*. doi:10.1016/j.envdev.2020.100529.

McCloskey, M. (1998). 'The Emperor Has No Clothes: The Conundrum of Sustainable Development'. *Duke Envtl. L. & Pol'y F.*, 9, 153.

McConney, P., Monnereau, I., Simmons, B. and Mahon, R. (2016). *Report on the Survey of National Intersectoral Coordination Mechanisms*. Barbados: The University of the West Indies, Centre for Resource Management and Environmental Studies. www.iwlearn.net/resolveuid/20f0c82c-6860-4fec-ac41-936a7ccac71f.

Navarre, M. and Lammens, H. (2017). 'Opportunities of Deep-Sea Mining and ESG Risks'. *Amundi Discussion Papers Series*, DP-24–2017. https://research-center.amundi.com/page/Publications/Discussion-Paper/2017/Opportunities-of-deep-sea-mining-and-ESG-ri.

Nwapi, C. and Wilde, D. (forthcoming). 'Enacting National Seabed Mineral Legislation in Africa for Areas beyond National Jurisdiction: Critical Issues for Consideration'. *Journal of Ocean Law and Governance in Africa*.

Okafor-Yarwood, I., Kadagi, N. I., Miranda, N. A., Uku, J., Elegbede, I. O. and Adewumi, I. J. (2020). 'The Blue Economy–Cultural Livelihood–Ecosystem Conservation Triangle: The African Experience'. *Frontiers in Marine Science*, 7, 586. doi:doi:10.3389/fmars.2020.00586.

Organisation for Economic Co-operation and Development (OECD) (2016). *The Ocean Economy in 2030*. Paris: OECD Publishing. https://doi.org/10.1787/9789264251724-en.

Patil, P., Virdin, J., Diez, S., Roberts, J. and Singh, A. (2016). *Toward a Blue Economy: A Promise for Sustainable Growth in the Caribbean*. Washington, DC: World Bank.

Pew Charitable Trusts (2020). *The Cape Town Agreement Explained.* www.pewtrusts.org/en/research-and-analysis/issue-briefs/2018/10/the-cape-town-agreement-explained (accessed 19 November 2020).

Scheld, A., Bilkovic, D. and Havens, K. (2016). 'The Dilemma of Derelict Gear'. *Sci Rep*, 6. doi:doi:10.1038/srep19671.

Sherman, K. and Alexander, L. (1986). *Variability and Management of Large Marine Ecosystems.* Boulder, CO: Westview Press.

Tolba, M. (2008). *Global Environmental Diplomacy: Negotiating Environmental Agreements for the World, 1973–1992.* Cambridge, MA: MIT Press.

United Nations (UN) (1992). *Rio Declaration on Environment and Development.* Vol. 1, Annex I, Principle 15. New York: United Nations.

United Nations (UN) (2020). *Progress towards the Sustainable Development Goals: Report of the Secretary-General.* New York: UN Economic and Social Council.

United Nations Conference on Trade and Development (UNCTAD) (2014). *The Oceans Economy: Opportunities and Challenges for Small Island Developing States.* New York: United Nations Conference on Trade and Development.

United Nations Economic Commission for Africa (UNECA) (1988). *Report of The Third Regional Conference on the Development and Utilisation of Mineral Resources in Africa, Kampala, Uganda, 6–15 June 1988, Annex II.* Addis Ababa: United Nations Economic Commission for Africa. https://repository.uneca.org/bitstream/handle/10855/5709/Bib-44970.pdf?sequence=1&isAllowed=y.

United Nations Economic Commission for Africa (UNECA) (2016). *Africa's Blue Economy: A Policy Handbook.* Addis Ababa: United Nations Economic Commission for Africa.

United Nations Environmental Programme (UNEP) (2015). *Blue Economy: Sharing Success Stories to Inspire Change.* Nairobi: United Nations Environmental Programme Regional Seas Report and Studies.

United Nations Environmental Programme (UNEP) (2019). *Environmental Rule of Law: First Global Report.* Nairobi: United Nations Environment Programme.

United Nations Environmental Programme (UNEP) (2020a). *Enabling Sustainable, Resilient and Inclusive Blue Economies.* www.unenvironment.org/explore-topics/oceans-seas/what-we-do/enabling-sustainable-resilient-and-inclusive-blue-economies (accessed 19 November 2020).

United Nations Environmental Programme (UNEP) (2020b). *Eastern Africa Region.* www.unenvironment.org/explore-topics/oceans-seas/what-we-do/working-regional-seas/regional-seas-programmes/eastern-africa (accessed 19 November 2020).

Van Nijen, K., Van Passel, S., Brown, C. G., Lodge, M. W., Segerson, K. and Squires, D. (2019). 'The Development of a Payment Regime for Deep Sea Mining Activities in the Area through Stakeholder Participation'. *International Journal of Marine and Coastal Law*, 34, 571–601. doi:doi:10.1163/15718085-13441100.

World Bank (2017). *The Sunken Billions Revisited: Progress and Challenges in Global Marine Fisheries.* Washington, DC: World Bank. doi:doi:10.1596/978-1-4648-0919-4.

World Bank (2019). *This Is What It's All About: Building Resilience and Adapting to Climate Change in Africa, March 7.* Washington, DC: World Bank.

Zacharias, M. and Ardron, J. (2020). *Marine Policy: An Introduction to Governance and International Law of the Oceans*, 2nd edn. New York: Routledge.

13 Making the blue economy happen

Clever Mafuta, MSc

Introduction

The blue economy has the potential to make a significant contribution to national economies in sub-Saharan Africa because it entails traditional maritime activities such as fisheries, tourism, mining, boat building, navigation, shipping and ports as well as new industries such as blue carbon (the term for carbon captured by the world's ocean and coastal ecosystems such as mangroves, seagrasses and saltmarshes), aquaculture, renewable energy (including wind, wave and tidal energy), bioproducts (such as pharmaceuticals and agrichemicals) and desalination.[1]

There are scholars who are of the view that the blue economy is both fluid and opaque, and that it can be described in different terms, including blue growth, ocean enterprise and sustainable ocean economy.[2] Furthermore, they argue that giving the term some shades of blue shows that the blue economy is a good business use of natural capital to support livelihoods and drive innovation.

The World Bank states that the blue economy is made up of various economic sectors and policies that ensure that the use of oceanic resources is sustainable.[3] Such sustainability is made possible through management practices that encompass sustainable fisheries, marine ecosystem health and prevention of pollution.

An overview of sub-Saharan Africa and the blue economy

According to the Organisation for Economic Co-operation and Development (OECD), the global value of the blue economy is currently estimated at US $1.5 trillion and employs 31 million people.[4] The OECD has forecast that by 2030 the value of the blue economy is set to double, to $3 trillion, spurred by aquaculture, offshore wind energy, fisheries and shipbuilding.

The importance of the blue economy to Africa is not only explained by the large ratio of the continent's population that lives within 100 km of the coastline and the economic activities that they are engaged in, but also by the large volume of international trade that is conducted through both oceanic and freshwater waterways. More than 90% of Africa's international trade is conducted by sea, while ocean fish contributes to the food and nutritional needs of millions of people on the continent, as well as providing an income to nearly 10 million people.[5]

The use of the oceans and seas in various economic activities is not only expanding but it is also being impacted by climate change. In addition, the growing and urbanizing coastal and riverine population is putting pressure on marine resources, resulting in environmental changes. Such changes affect Africa's blue economy, and there is a need for a suite of global, regional and national policies to facilitate sustainable development.

As discussed in previous chapters in this volume, the need for policies to ensure that countries reap the maximum benefits from blue economic activities is further justified by

> threats such as piracy and armed robbery, the trafficking of people, illicit narcotics and weapons, as well as natural threats from tsunamis and hurricanes, and rising sea levels and ocean acidification. Overfishing caused by illegal, unreported, and unregulated (IUU) fishing and other unsustainable fishing practices also pose a serious problem in the region, along with pollution and habitat destruction.[6]

While the blue economy is considered to be important for island nations such as Mauritius, Seychelles, Comoros, Madagascar, Cabo Verde, and São Tomé and Príncipe, even mainland coastal countries derive huge benefits from blue economic activities. For example, as shown in Chapter 7 in this volume and as will be discussed in this chapter, South Africa's Operation Phakisa seeks to bolster the country's economic and social gains through making use of the ocean resources. Operation Phakisa was launched in 2014 and sought to increase South Africa's annual ocean economy contribution from US $3.7 billion to $11.5 billion while also creating 800,000–1 million jobs by 2030.[7]

Appropriate policies for sustainable management

Effective policies for the blue economy need to be grounded in sound information and science. Various tools are available to inform policymaking, including marine spatial planning (MSP) and ecosystem-based management (EBM).

Marine spatial planning

South Africa's national framework for MSP describes it as the governance process to assess and manage the distribution of human activities to achieve economic, social and ecological objectives in both space and time. The process of such planning is area-based by targeting the marine space that people can use and care for. The planning is also integrated across many sectors of the economy, and should aim to maximize benefits in a sustainable way.[8]

The application of MSP enables countries to obtain economic, social, ecological and governance benefits while allowing for sustainable development. Traditionally, MSP was limited to managing conflicts between competing human uses in their ocean space. MSP is increasingly being introduced in emerging economies to facilitate better planning of activities and management of resources over time at both the country and regional level. For example, South Africa works with its neighbours under the Benguela Current Commission in developing the necessary capacities to pursue MSP.[9]

MSP is key to achieving policy goals designed to unlock the ocean economy, ensure healthy marine ecosystems, and contribute to good ocean governance. Through MSP littoral states are better able to exercise their sovereign power over coastal sea spaces than they would have been had no such planning been in place.[10]

Ecosystem-based management

Another approach used to guide blue economy policymaking is the adoption of ecosystem-based management (EBM). According to Grieve and Short, the major goal of EBM is to exploit natural resources sustainably by acknowledging the effect of the environment on those resources, as well as acknowledging the effect of their exploitation on the environment.[11] As noted by Grieve and Short, the objectives of EBM are:

- to maintain the natural structure and function of ecosystems and their productivity;
- to incorporate uses and values of ecosystems in management approaches; and
- to recognize that ecosystems are dynamic and always changing, and hence the need for scientific knowledge as part of continuous learning.

Ward suggests that the identification of stakeholders should be one of the first steps in EBM approaches for the fisheries sector. Stakeholder involvement facilitates transparency and accountability.[12] Also of importance in EBM in the fisheries sector is the mapping of important ecosystems and habitats. Such a two-step process gave policy direction to the Rufiji, Mafia and Kilwa districts of Tanzania insofar as collaborative engagement, knowledge sharing, enterprise development, and protection of threatened habitats and species were concerned.[13]

Global and regional policies

The blue economy comprises a number of different sectors, which necessitates the development of integrated legal, regulatory and institutional frameworks. However, much of sub-Saharan Africa is characterized by policy incoherence, weak enforcement, and policy gaps which indicates that the current legislative framework is fragile.

The multisectoral nature of the blue economy comes at a time when there are increasing socio-economic demands in a world faced by unprecedented challenges such as the coronavirus (COVID-19) pandemic and the resultant global economic recession. As such, the world is being forced to seek new opportunities for creating wealth, including making greater use of resources from water bodies. Given their vast nature, world attention has been turning to the political economy of the sea, resulting in the development of global policies such as the United Nations Convention on the Law of the Sea (UNCLOS). The UN Sustainable Development Goals (SDGs) have also set targets on the wise use of ocean resources: for example, the goal of SDG 14.7 is to 'By 2030, increase the economic benefits to small island developing States and least developed countries from the sustainable use of marine resources, including through sustainable management of fisheries, aquaculture and tourism'.[14]

As discussed in the previous chapter, UNCLOS, which came into force in 1996, offered coastal states large swathes of sea space to explore and exploit. The UNCLOS framework law facilitated MSP which resulted in countries being able to delineate their maritime boundaries and extend their continental shelves. As a result, island nations such as Cabo Verde, Comoros, Madagascar, Mauritius, São Tomé and Príncipe, and Seychelles gained control of sea space, which in the majority of cases far exceeds their land area.[15] De Vivero and Matoes reported that 58.42% of world sea space fell under national jurisdiction after UNCLOS came into operation, resulting in some tension among countries given the earlier understanding of the sea as a common heritage of humanity.[16]

In addition to allowing countries to delineate their maritime boundaries and extend their continental shelves, the UNCLOS framework law also explains the rights of states with regard to navigation, access to living and non-living resources, protection and preservation of the marine environment, marine scientific research, and development and transfer of marine technology. The Convention also seeks to strengthen peace, security, cooperation and friendly relations among all nations.[17]

There are numerous international conventions, many of which are relevant to the growth of sub-Saharan Africa's blue economy. The international instruments include the Convention on Wetlands of International Importance (Ramsar Convention) and the Convention on Biological Diversity both of which are key to the protection of biodiversity and to the securing of economic value from such resources. Other conventions that support the blue economy are the Convention for the Protection, Management and Development of the Marine and Coastal Environment of the Eastern African Region (Nairobi Convention); the Convention for Co-operation in the Protection and Development of the Marine and Coastal Environment of the West and

Central African Region (Abidjan Convention); the International Convention for the Prevention of Pollution from Ships (MARPOL); the African Maritime Transport Charter; the African Convention on the Conservation of Nature and Natural Resources; and the World Heritage Convention. The World Trade Organization is also key to Africa's trade and economic activities that also support the continent's blue economy.[18]

While global policy pronouncements such as the SDGs and UNCLOS are key to informing blue economy policies, regional groupings such as the African Union (AU) are also important in setting the continental policy agenda. The AU is building an Africa-wide consensus on the role that the blue economy could play in the transformation of Africa. The AU's 2050 Africa's Integrated Maritime Strategy (2050 AIM Strategy) describes the blue economy as the 'new frontier of the African renaissance'.[19] Furthermore, the blue economy is addressed in the AU's Agenda 2063, which recognizes that the blue economy can be a catalyst for Africa's socio-economic transformation.[20]

In 2012, at its 13th Ordinary Session, the AU Assembly called upon the AU Commission to develop the 2050 Africa's Integrated Maritime Strategy (2050 AIM Strategy).[21] The Strategy is based on the need to address illegal activities such as toxic waste dumping and illegal discharge of oil, dealing in illicit crude oil, arms and drug trafficking, human trafficking and smuggling, piracy and armed robbery at sea; as well as to explore green growth opportunities in energy exploitation, climate change, environmental protection and conservation and safety of life and property at sea. The strategy also seeks to advance maritime research, innovation and development, with the ultimate goal of enhancing Africa's maritime competitiveness, job creation, international trade, maritime infrastructure, transport, information, and improved communication, technology and logistics.

The 2050 AIM Strategy was developed to foster greater wealth creation from Africa's oceans, seas and inland waterways. The strategy has a plan of action that outlines activities, outcomes, time frames and executing agents.

While there are some policy pronouncements on the blue economy at the AU level, these policies are often not supported by adequate institutional mechanisms. The AU's technical arms such as the African Ministerial Council on Water and the African Ministerial Conference on the Environment are hardly mandated to facilitate regional collaboration on blue economy.

National blue economy policies

Despite the weak institutional arrangements, there are efforts to have blue economy policy arrangements even at subcontinental levels. According to the UN Economic Commission for Africa (UNECA), regional economic communities and states are increasingly recognizing the importance of the blue economy by developing blue economy strategies.[22] The Indian Ocean

Commission is developing a Blue Economy Action Plan for its members, while at the country level blue economy strategies are being pursued in many countries including Mauritius, Seychelles, Senegal and South Africa.

Mauritius has a policy framework which was developed through broad-based consultations with all stakeholders, and this created a sense of ownership in both the process and the product, as well as allowing for accountability which is key in both the implementation and achievement of goals and aspirations. The country also enacted relevant laws to give effect to the blue economy policy.[23]

As shown in Chapter 8 in this volume, Seychelles has a dedicated government body to promote the blue economy through the harnessing of the country's ocean resources for not only achieving sustainable development, but also combating climate change. The Department of the Blue Economy in the Ministry of Finance, Trade, Investment and Economic Planning superintends the implementation of the blue economy in Seychelles by creating conditions for better coordination and cooperation with other sectors of the economy. The country also has a blue economy roadmap.[24]

In 2006 Senegal created a coordination mechanism, the Haute Autorité chargéé de la coordination de la sécurité maritime et de la protection de l'environnement marin (HASSMAR), to address maritime security issues and the protection of the marine environment. Through HASSMAR, maritime agencies work together in national planning and interventions at sea.[25]

As stated earlier, South Africa launched its Operation Phakisa initiative in 2014 as a framework policy for the creation of development and wealth from South Africa's blue economy. Operation Phakisa prioritizes transport and manufacturing, offshore oil and gas, aquaculture, and protection and governance as sectors of the country's ocean growth. The initiative is run out of the country's office of the president. The transport and manufacturing sectors priority area aims to gain benefits that can be accrued from storage and warehousing, ship and boat building, and rig repairs. The offshore oil and gas sector seeks to create opportunities for exploration through enabling policy and legislation, inclusive economic growth, and developing infrastructure. The aquaculture sector seeks to develop business and employment opportunities in fish processing and marketing.[26]

Conclusion and recommendations

The majority of sub-Saharan African countries are signatories to UNCLOS. While this is a plausible for national policy direction, it is also worth noting that some of the countries are yet to domesticate the provisions of UNCLOS into their national policies and laws. In addition, many countries in the region have yet to make full use of the provisions of the two key instruments under UNCLOS with regards to the sustainable exploitation of resources in the exclusive economic zones (EEZ), and the exploration and exploitation of resources in the common heritage area. In order for sub-Saharan Africa to grow its blue economy, not only do countries need to sign up to UNCLOS but to domesticate the law, and to fully embrace the instruments that give them access to resources in the common heritage areas and in their EEZ.

Tools such as MSP and EBM are increasingly being used to guide regional and national policy. While island nations such as Seychelles and Mauritius have managed to extend their national areas of jurisdiction beyond the EEZ as provided for under UNCLOS and as enabled by tools such as MSP, many coastal countries in sub-Saharan Africa have yet to expand their area of jurisdiction and to reap the full benefits of the blue economy.

Notes

1 World Bank and United Nations Department of Economic and Social Affairs. (2017). *The Potential of the Blue Economy: Increasing Long-Term Benefits of the Sustainable Use of Marine Resources for Small Island Developing States and Coastal Least Developed Countries.* Washington, DC: World Bank.
2 Voyer, M., Quirk, G. and McIlgorm Azmi, K. (2018). 'Shades of Blue: What Do Competing Interpretations of the Blue Economy Mean for Oceans Governance?' *Journal of Environmental Policy & Planning, 20*(5), 595–616. doi:10.1080/1523908X.2018.1473153.
3 World Bank and United Nations Department of Economic and Social Affairs. *Potential of the Blue Economy.*
4 Organisation for Economic Co-operation and Development (OECD) (2016). *The Ocean Economy in 2030.* Paris: OECD Publishing.
5 African Union (AU) (2012). *2050 Africa's Integrated Maritime Strategy (2050 AIM Strategy).* https://cggrps.com/wp-content/uploads/2050-AIM-Strategy_EN.pdf (accessed 25 November 2020).
6 United Nations Economic Commission for Africa (UNECA) (2016). *Africa's Blue Economy Handbook.* https://gridarendal-website-live.s3.amazonaws.com/production/documents/:s_document/329/original/blueeco-policy-handbook_en.pdf?1490620941 (accessed 23 November 2020).
7 Findlay, K. and Bohler-Muller, N. (2018). 'South Africa's Ocean Economy and Operation Phakisa: Lessons Learned'. In V. N. Attri and N. Bohler-Muller (eds), *The Blue Economy Handbook of the Indian Ocean Region* (pp. 231–55). Pretoria: Africa Institute of South Africa.
8 United Nations Department of Environmental Affairs (2017). https://iwlearn.net/resolveuid/21f3aa17-e74c-4f3c-8682-9e28af296b20 (accessed 23 November 2020).
9 Ibid.
10 Maes, F. (2008). 'The International Legal Framework for Marine Spatial Planning'. *Marine Policy,* Vol. 32, pp. 797–810.
11 Grieve, C. and Short, K. (2007). *Implementation of Ecosystem-Based Management in Marine Capture Fisheries.* https://wwfeu.awsassets.panda.org/downloads/wwf_ebm_toolkit_2007.pdf (accessed 23 November 2020).
12 Ward, T., Tarte, D., Hergel, E. and Short, K. (2002). *Policy Proposals and Operational Guidance for Ecosystem-Based Management of Marine Capture Fisheries.* Sydney: WWF; Maes, 'International Legal Framework for Marine Spatial Planning'.
13 Grieve and Short, *Implementation of Ecosystem-Based Management in Marine Capture Fisheries.*
14 United Nations Development Programme (UNDP) (2020). *Goal 14 Targets.* www.undp.org/content/undp/en/home/sustainable-development-goals/goal-14-life-below-water/targets.html (accessed 24 November 2020).

15 Forbes, V. (1995). *The Maritime Boundaries of the Indian Ocean Region.* Singapore: Singapore University Press.
16 Suarez de Vivero, J. and Rodriguez Mateos, J. (2010). 'The BRIC Countries as Emerging Powers: Building New Geopolitical Scenarios'. *Marine Policy*, Vol. 34, pp. 967–78.
17 UNECA, *Africa's Blue Economy.*
18 Ibid.
19 African Union (AU) (2015). *For the Launch of the 2015–2025 Decade of African Seas and Oceans and the Celebration of the African Day of the Seas and Oceans on 25 July 2015.* Addis Ababa: AU.
20 African Union (AU) (2015). *22nd Ordinary Session of AU Heads of State and Governments on the Adoption and Implementation of the AU 2050 AIMS.*www.au.int/en/sites/default/files/decisions/9659-assembly_au_dec_490–516_xxii_e.pdf (accessed 24 November 2020).
21 African Union, *2050 Africa's Integrated Maritime Strategy.*
22 UNECA, *Africa's Blue Economy.*
23 United Nations Conference on Trade and Development (UNCTAD) (2014). *The Oceans Economy: Opportunities and Challenges for Small Island Developing States.*https://unctad.org/system/files/official-document/ditcted2014d5_en.pdf (accessed 25 November 2020).
24 Ministry of Foreign Affairs and Tourism (2015). *First Consultation by new Minister of Finance, Trade and the Blue Economy with Chamber of Commerce.* www.mfa.gov.sc/static.php?content_id=36&news_id=967 (accessed 25 November 2020).
25 Government of Senegal (2006). *Republique du Senegal Primature: Journal Officiel.* www.jo.gouv.sn/spip.php?article4968 (accessed 25 November 2020).
26 Department of Planning, Monitoring and Evaluation of the Republic of South Africa (2014). *Operation Phakisa.*www.operationphakisa.gov.za/Pages/Home.aspx (accessed 25 November 2020).

Afterword

What is possible? What is likely?

Donald L. Sparks, PhD

The contributions in this volume show the enormous challenges and opportunities the blue economy presents for sub-Saharan Africa. Clearly the largest current challenge is the coronavirus (COVID-19) pandemic. While its effects are difficult to assess at this point in time, the costs in terms of human lives and the global economy have been horrendous. The United Nations Department of Economic and Social Affairs recently looked at the effect that the pandemic has had on ocean acidification. Prior to 2019 it predicted that ocean acidity would increase by 125%, and this would affect half of all marine life. The Department now suggests that the drastic reduction in human activity as a result of COVID-19 may be a chance for the ocean to recuperate. Perhaps that is a silver lining in this very dark cloud.

The blue economy shows much promise, including finding ways to pursue a low carbon path of economic development that would include creating employment opportunities and reducing poverty. Indeed, the World Bank believes that 'Blue growth, or environmentally sustainable economic growth based on the oceans, is a strategy of sustaining economic growth and job creation necessary to reduce poverty in the face of worsening resource constraints and climate crisis'.[1]

However, the blue economy also faces numerous challenges. Until recently, the world's coastal and marine areas have been thought of as having limitless resources and being somewhere to store our waste. The results of man's misuse of the global water resources range from degraded coastal habitats and marine pollution to the negative impacts of man-made climate change and overfishing. According to the Food and Agriculture Organization of the UN (FAO), some 57% of the world's fish stocks are fully exploited and another 30% are over-exploited.[2] Marine fisheries generally contribute some US $270 billion annually to the world economy, according to FAO.

In addition to the valuable contributions in this volume, several other recent studies have been conducted by the African Union (AU), the World Bank and the UN Economic Commission for Africa (UNECA) that look at the future of the blue economy in sub-Saharan Africa.

The AU cites a number of continent-wide initiatives to move the blue economy objectives forward.[3] They include:

- The African Union's Agenda 2063. This sets the strategic goal of economic transformation including specific goals for the blue and ocean economy as goal number 6 (maritime resources and energy) and goal number 7 (port operations and marine transportation). Other related priorities include sustainable natural resource management, renewal energy and water security.
- The 2050 Africa's Integrated Maritime Strategy (2050 AIM Strategy). The strategy prioritizes marine conservation, research, education and governance to address maritime challenges.
- The 2014 Policy Framework and Reform Strategy for Fisheries and Aquaculture in Africa. The Framework will help African governments to develop appropriate fisheries and aquaculture policies; and
- The 2016 African Charter on Maritime Security and Safety and Development in Africa (the Lomé Charter). The Charter aims to prevent national and transnational crime such as piracy and all forms of illegal trafficking.

As discussed in Chapter 12 in this volume, in addition to these continent-wide efforts there are a number of regional initiatives currently underway under the direction of the Indian Ocean Commission, the Southern African Development Community and the Indian Ocean Rim Association.

UNECA has also suggested a number of challenges and opportunities for the region.[4] The former include rising ocean temperatures and ocean acidification resulting in a further decline in fish stocks and the loss of livelihoods that depend on them. On the positive side, the opportunities include low-carbon technological innovations that shift mindsets towards a better relationship with the natural world. This includes marine energy development such as wave energy (currently being developed in Cabo Verde) and tidal power. UNECA's policy handbook appealed for the mainstreaming of climate change and environmental sustainability into realistic national and regional policies; increasing investments in climate information services; strengthening environmentally sustainable infrastructure (e.g. 'green' ports and renewable energy); linking ocean energy with high-value economic activities such as tourism; implementing better environmental impact assessments and strategic environmental assessments in the blue economy; enhancing early warning systems to mitigate ocean-related risks; assisting with the enhancement of national capacities to better cooperate with international agencies such as the International Union for the Conservation of Nature; helping to expand and/or initiate marine protected areas; raising awareness of these related issues within the region; helping to devise national legislation to promote blue economies in selected countries; creating a national capital accounting system that will allow states to promote sustainable energy (and carbon taxes to help to finance this); and harmonizing regional and continental approaches within the blue economy context.

In its 2017 blue economy study the World Bank identified pathways that countries could take for future action.[5] While realizing there are national and local priorities that may conflict, the study identified 10 common steps which include:

1. More accurately value natural resource capital including trade-offs among different blue economy sectors;
2. Invest in (and use of) the best science, data and technology;
3. Countries should individually weigh up the relative importance of their blue economy sectors to decide priorities;
4. Anticipate and adapt to climate change;
5. Adopt new investment techniques including blue bonds, debt-for-adaption swaps and other innovative instruments;
6. Fully implement the UN Convention on the Law of the Sea;
7. Invite the active participation of all societal groups, especially women, young people, indigenous peoples and under-represented groups;
8. Develop coastal and marine spatial plans to guide better decision-making for the blue economy;
9. Encourage the private sector to play a key role; and,
10. Find effective partnerships especially from Small Island Developing States and coastal least developed countries.

Several contributors to this volume identified some specific challenges and opportunities that lie on the horizon.[6]

Governance

From a governance perspective, states must work towards integrated marine policies that facilitate blue growth, and address sustainability issues to avoid acute and cumulative impacts. These policies must be translated into laws that set standards and provide clear regulatory frameworks for expanded ocean industries. Administrative structures and decision-making processes must also support sustainability. Such reforms may be possible in more developed well-resourced countries. For many nations, however, a lack of funding and technical expertise, as well as political will, may hamper, or at least delay, the achievement of these goals. Perhaps Hardman-Mountford puts it best in Chapter 12: ‘Ultimately, the touchstone of success in implementing the blue economy must not be solely the year’s economic returns, but rather the rate of transition from unsustainable to sustainable activities that will yield returns over the much longer term’.

The sea is inherent to the complex story of sub-Saharan Africa. Maritime interaction with the rest of the world substantially contributed to forging the unique maritime culture of the region and also influenced its socio-economic character over a long period of time. Africa’s access to the sea and its own wealth made it both receptive and vulnerable to what the sea brought. Since ancient times, contact with Arabian and Eastern civilizations, Europe and beyond, has influenced the region's mercantile, social and political history.

Sub-Saharan Africa's initial commercial and coastal partnerships were linked to a thirst for gold, ivory and slaves. Over time these evolved into complex and prosperous maritime trade and extensive social-cultural interaction, which eventually made way for resistance and colonial conquest. During the second half of the twentieth century this was followed by nationalism, triumph, freedom and realignment. In an environment characterized by a quest for raw materials, it was necessary for Africa to re-establish its political and commercial relationship with the rest of world.

The maritime interests of the region are considerable: its oceans and ports are inherent to the wealth of the region; they are crucial for trade; and they are an important source of food and energy. However, regional governments do not give sufficient attention to the blue economy and its potential to stimulate sustainable economic growth and development. The lesson of how important sub-Saharan Africa's close historical association with the oceans once was should be relearned. Its maritime legacy must be reinforced with strong international and regional commercial cooperation, good governance, environmental care, maritime security and sound business approaches to contribute to the collective charting of a sustainable development course for the region.

South Africa

At a country-specific level, South Africa's large sea area, abundant living and non-living marine resources, considerable maritime infrastructure and ocean-related economic activities contribute substantially to the country's gross domestic product and are crucial to its socio-economic growth, development and transformation. What could be called the South African blue economy approach is an important focus area for the South African government and forms part of an ambitious ocean economy programme under Operation Phakisa.

It is vital that the exploitation of maritime resources is economically sustainable, well managed, and aimed at the preservation of fragile marine ecosystems. This is specifically pertinent as the illicit benefits derived from the oceans have proven to be both very profitable and damaging to the country's ocean ecosystem. With fish stocks under considerable pressure, the situation is dire, and at its current levels illegal, unreported and unregulated fishing and poaching will result in certain species ceasing to exist in the wild. Other facets of the South African blue economy, from oil and gas initiatives, infrastructure improvements, to enhanced trade and shipbuilding, are a mixed bag of opportunities, growth and frustrations. Unfortunately, coastal and marine tourism, an important growth segment of the South African economy, have been adversely affected by COVID-19.

South Africa is poised for enhanced blue economy growth and development due to its infrastructure, abundant resources and geographic location. The provisos to this success, however, include investor confidence, maritime security, good governance and environmental sustainability.

Energy

There is no doubt that the offshore oil and gas industry was disrupted by COVID-19 during 2020. Oil prices were (and remain at the time of writing) volatile, demand is down, supply is up, and storage capacity is limited. In such turbulent times, there exists a need to act swiftly in order to keep the business viable, to come back strong when conditions improve. Companies that invest in new technology today will be ahead of the competition when the industry recovers. Digitalization of the value chain, also known as value chain optimization, is an essential part of staying competitive.

Marine mineral mining has been carried out for many years, with most commercial ventures focusing on aggregates, diamonds, tin, magnesium, salt, sulphur, gold and heavy minerals. Activities have generally been confined to the shallows close to the shore (in water less than 50 m deep), but the industry is evolving and mining in deeper water looks set to proceed, with phosphate, massive sulphide deposits, manganese nodules and cobalt-rich crusts regarded as potential future prospects. Seabed mining is a relatively small industry and involves only a fraction of the known deposits of marine minerals.

Ports

As a generalization, for sub-Saharan African port operations, consolidation of container handling at the site will improve efficiency and vessel turnaround times. Modified existing terminals and new terminals will result in greater economies of scale, optimize the deployment of resources for port and marine services by automating both wharf-side and yard operations, and reduce the need for inter-terminal haulage. Infrastructure for intra- and interstate transportation is required for a smooth, seamless transition of cargo from ship to point of destination and vice versa.

Local restrictions on price increases and the regional need to remain competitive in sub-Saharan Africa suggests that shippers are unlikely to have to pay more when cargo operations start at new terminals. Innovations should be piloted for ports – small, medium and large – and should include next-generation vessel traffic management systems that are able to predict congestion hotspots and assist vessel route planning. Trialling of a maritime 'single window' that improves port clearance (the concept of digitalization) by streamlining the submission process and a 'just-in-time' planning and coordination system for faster vessel turnarounds are other possibilities. Remote-controlled vessel pilotage and a maritime sense-making system to boost port operations by preventing illegal bunkering and optimizing anchorage utilization are among other innovations under consideration. Development of the first phase of intelligent and innovative systems, including a 'smart grid' with sustainability in mind, is essential.

Conclusion

Expanded economic growth and development is promoted by the business community, economists and politicians; however, it may well not be in humankind's best long-term interest for the planet and the international community. Our oceans are stressed, coral is dying, waters are warming due to climate change and species are not recovering from these events, never mind overfishing and pollution of the marine environment. We need to value our oceans and understand them better before we inflict more damage on them. The ocean contains multiple ecosystems; hence, we cannot dismiss working in all of them because one or two are overstressed. There must be a balance between which resources are taken from the ocean and utilized by humankind and the need to make a profit solely to benefit a specific business or industry or a particular political agenda at the global, regional, national and local government level.

The influence of blue economy policies and goals is likely to continue to expand. Just as sustainable development initiatives have flourished, so too will blue economy projects and activities. It is likely that all states will adopt strategies to achieve enhanced wealth and well-being from the oceans in the coming years. Whether blue economy outcomes will live up to their promise is another matter. There is little doubt that short-term gains will be made including from increased aquaculture and advances in fishing technologies, for example. Yet the long-term sustainability of these endeavours will require concerted effort and significant political will.

In conclusion, Attri and Bohler-Muller summarize it best:

> Paradoxically, the fundamental challenge of the blue economy is simultaneously to do more and to do less. Put simply, the blue economy should increase the overall wealth of nations derived from ocean and coastal resources, although increased wealth in fisheries (for instance) may require reducing fishing to biologically sustainable levels.[7]

To succeed, each country must chart a course that is appropriate to its development aspirations, keeping in mind that such efforts must be completed in a regionally coordinated manner.

Notes

1 World Bank. *The Potential of the Blue Economy*. Washington, DC: World Bank, 2017, p. 5.

2 Food and Agriculture Organization of the United Nations (FAO). 'The State of the World's Fisheries and Aquaculture 2016: Contributing to Food Security and Nutrition for All'. Rome: FAO, 2016, p. 12.
3 African Union. *Africa Blue Economy Strategy.* Nairobi: IBAR, 2019.
4 United Nations Economic Commission for Africa (UNECA). *Africa's Blue Economy: A Policy Handbook.* Addis Ababa: UNECA, 2016.
5 *The Potential of the Blue Economy.* World Bank.
6 I would like to acknowledge and thank my fellow contributors Viv Forbes, Erika Techera, Charles Colgan and Nick Hardman-Mountford.
7 Attri, Vishva Nath and Narnia Bohler-Muller. *The Blue Economy Handbook of the Indian Ocean Region.* Johannesburg: Africa Institute of South Africa, 2018.

Appendix: UN Sustainable Development Goal number 14 targets

Target 14.1

By 2025, prevent and significantly reduce marine pollution of all kinds, in particular from land-based activities, including marine debris and nutrient pollution.

Target 14.2

By 2020, sustainably manage and protect marine and coastal ecosystems to avoid significant adverse impacts, including by strengthening their resilience, and take action for their restoration in order to achieve healthy and productive oceans.

Target 14.3

Minimize and address the impacts of ocean acidification, including through enhanced scientific cooperation at all levels.

Target 14.4

By 2020, effectively regulate harvesting and end overfishing, illegal, unreported and unregulated fishing and destructive fishing practices and implement science-based management plans, in order to restore fish stocks in the shortest time feasible, at least to levels that can produce maximum sustainable yield as determined by their biological characteristics.

Target 14.5

By 2020, conserve at least 10 per cent of coastal and marine areas, consistent with national and international law and based on the best available scientific information.

Target 14.6

By 2020, prohibit certain forms of fisheries subsidies which contribute to overcapacity and overfishing, eliminate subsidies that contribute to illegal, unreported and unregulated fishing and refrain from introducing new such subsidies, recognizing that appropriate and effective special and differential treatment for developing and least developed countries should be an integral part of the World Trade Organization fisheries subsidies negotiation.

Target 14.7

By 2030, increase the economic benefits to small island developing States and least developed countries from the sustainable use of marine resources, including through sustainable management of fisheries, aquaculture and tourism

Target 14.a

Increase scientific knowledge, develop research capacity and transfer marine technology, taking into account the Intergovernmental Oceanographic Commission Criteria and Guidelines on the Transfer of Marine Technology, in order to improve ocean health and to enhance the contribution of marine biodiversity to the development of developing countries, in particular small island developing States and least developed countries.

Target 14.b

Provide access for small-scale artisanal fishers to marine resources and markets.

Target 14.c

Enhance the conservation and sustainable use of oceans and their resources by implementing international law as reflected in UNCLOS, which provides the legal framework for the conservation and sustainable use of oceans and their resources, as recalled in paragraph 158 of The Future We Want.

Source: https://sdgs.un.org/goals/goal14.

Index

Page numbers in *italics* and **bold** indicate Figures and Tables, respectively.

For Product Safety Concerns and Information please contact our EU representative GPSR@taylorandfrancis.com
Taylor & Francis Verlag GmbH, Kaufingerstraße 24, 80331 München, Germany

www.ingramcontent.com/pod-product-compliance
Lightning Source LLC
Chambersburg PA
CBHW052121230326
41598CB00080B/3938